C000296800

The Technology of Nonviolence

The Technology of Nonviolence

Social Media and Violence Prevention

Joseph G. Bock
Foreword by John Paul Lederach

The MIT Press
Cambridge, Massachusetts
London, England

© 2012 Massachusetts Institute of Technology

All rights reserved. No part of this book may be reproduced in any form by any electronic or mechanical means (including photocopying, recording, or information storage and retrieval) without permission in writing from the publisher.

MIT Press books may be purchased at special quantity discounts for business or sales promotional use. For information, please email special_sales@mitpress.mit .edu or write to Special Sales Department, The MIT Press, 55 Hayward Street, Cambridge, MA 02142.

This book was set in Stone Sans and Stone Serif by Toppan Best-set Premedia Limited. Printed and bound in the United States of America.

Library of Congress Cataloging-in-Publication Data

Bock, Joseph G.
The technology of nonviolence : social media and violence prevention / Joseph G. Bock; foreword by John Paul Lederach.
    p. cm.
Includes bibliographical references and index.
ISBN 978-0-262-01762-6 (hardcover : alk. paper)
1. Nonviolence. 2. Violence—Prevention. 3. Social media. I. Title.
HM1281.B63 2012
303.6′1—dc23
2011047646

10 9 8 7 6 5 4 3 2 1

Grateful acknowledgment is made for copyright permissions that have been granted by

• Wesley G. Skogan, Northwestern University's Institute for Policy Research, for permission to reproduce graphics presented in a 2008 evaluation of CeaseFire. They are reproduced herein as figures 1.3, 4.1, 4.2, and 4.3.

• Patrick Meier, Stanford University's Program on Liberation Technology and Ushahidi, for permission to reproduce his graphic depiction of the accuracy of early warning data over time. It is reproduced herein as figure 8.1.

• Susanne Schmeidl, for permission to quote her paper delivered at the International Studies Association Conference in 2008. It is quoted herein in chapter 12, "Future Directions and Recommendations."

Dedicated to
all those who risk their lives to save the lives of others
without feeling the security, however false that might be, of a gun in their hands;
and to my father
James Valentine Bock
who always helps me keep my rudder in the water.

The definition of the Arab Spring is increasingly being stretched. It's both about the current showdowns and the long-term spillover. The upheavals—supercharged by the instant communications of the Web—have given the region a crash course in the clout of the streets. The view from the top is suddenly less comfortable.

—"Arab Spring Hardens into Summer of Stalemates as Challenge of Changing Regimes Becomes Clearer"

There is much that aid organizations can do to build on the strategies that communities employ . . . [in order to] maintain their assets, escape violence, and mitigate threats.

—Sorcha O'Callaghan and Sara Pantuliano, "Protective Action: Incorporating Civilian Protection into Humanitarian Response"

Contents

Foreword   ix

Preface   xi

Acknowledgments   xix

Introduction   1

I   Theory and Methodology   15

1   Toward an Applied Theory of Violence Prevention   17

2   Reporting and Warning about Deadly Possibilities   37

II   Violence Prevention on the Ground   55

3   Organizing against Ethnoreligious Violence in Ahmedabad   57

4   Interrupting Gang Violence in Chicago   81

5   Counteracting Ethnoreligious Violence in Sri Lanka   91

6   Crowdsourcing during Post-election Violence in Kenya   105

7   Circumventing Tribal Violence in East Africa   127

8   Comparing the Approaches   135

9   How to Intervene Effectively   147

10   What to Do When Violence Prevention Is Unlikely to Work   161

III   Resource Allocation Considerations and Recommendations   177

11   Concerns about Misallocation of Resources   179

12   Future Directions and Recommendations   189

Conclusion   203

Appendix A: Reporting Sheet for Field Officers   209

Appendix B: Categories for Local Conflict Early Warning and Early Response   211

Appendix C: "Super Event" Categories   217

Appendix D: Indicators of the CEWARN Mechanism   221

Appendix E: Results from Statistical Analysis on Organized Raids   225

Acronyms   227
Glossary   231
Notes   239
References   257
Index   275

Foreword

The wider peacebuilding field has long suffered under the weight of an intriguing paradox: Violence has never been more studied, understood in terms of its dynamics and patterns, and officially identified as a policy challenge than it has in past the twenty years. And violence continues to unfold with extraordinary resilience in too many communities around the world.

In this book Joe Bock makes a significant step into unpacking and suggesting creative ways to respond to this dilemma. I say that as a practitioner-scholar with thirty years of experience in community conflict and as a critic of the difficulties that this promising area of exploration—all too often misnamed as conflict prevention—has failed to deliver. From the standpoint of the practice of strategic peacebuilding several key contributions are worth highlighting from this book.

First, let me elaborate on why this book has innovation. Creativity often involves mixing unexpected sources and ideas that break through toward new insights or understandings. Joe provides a service to our wider field by ranging significantly beyond the most commonly explored theories and approaches and linking them into a coherent, practical set of approaches. He covers wide-ranging theories from different fields of inquiry and practice, from technology studies to public health, from conflict studies to cutting-edge statistical methods. This improbable mix provides surprises and "ah-hah!" moments for the reader and practitioner. He consistently illustrates these ideas with concrete examples from real-life situations and learning. The net result provides greater clarity on understanding how violence prevention may improve and what may help shift our understanding toward responsive action.

Second, among the most important aspects of this book for the practitioner is its commitment to think from and within the context of the local community. Much of violence prevention has relied heavily on inputs from community sources but has sought responses from outside. Implicitly the local community has not been seen as a *resource* for response. We thus are often faced with a gulf that separates data collection, the cries of warnings coming from key people at local levels, and a lack of response from other levels of society. Generally speaking, the challenge of preventing, in essence, has always been how to move from early warning to early action. I think the key, suggested and supported by this book, is a shift in the exclusive focus from "early" toward appropriate and effective response. This requires not only a capacity for short-term responses but also the long and slow process of building platforms of response that include, inspire, and embolden local communities to envision themselves as having capacity for generating the signs and data of violence emergent and the resources to catalyze response. This book focuses as much on local community capacity and creativity as it does on the needed responses from mid- and higher-level policy and officials, and in so doing begins to fill the *warning-action gulf* with concrete suggestions and ideas that are practical and innovative.

Third, Joe takes seriously the old adage that there is nothing as practical as a good theory. This book more than any other I have seen fearlessly explores the theoretical side of practice, mining best examples and evidence-based research that provide us with something we have known for a long time but have not sufficiently applied: When good practice has clarity about its theory of change the combination provides far greater effective response and consistent learning. Peacebuilding and violence prevention are complex processes. They require us to bring our very best understanding and a commitment to continuous learning. This is one of the basic lessons proposed by this book. Best practices have clearly articulated theories of change that are in constant study and development.

Finally, this book comes at a very timely juncture. Those of us who have worked in areas of significant violence have increasingly been impressed that while these are often areas of extraordinary economic challenge, of social and psychological distress, they are also techno-savvy geographies. This is particularly true of the cell phone with its capacity for digital images and messaging. From Mindanao to Nepal to the recent events in the Arab Spring these popular, handheld and relatively low-cost accessible tools have created forms of social power with unequal historic comparison. Social movements are literally this: They *move* in timeframes and with mobilizing capacity that can make a real difference. What Joe documents and proposes is that this accessibility, recognition of social connection, and networking can join hands with the hopes of those people most affected on a daily basis by destructive conflict that violence can be faced, interrupted, and prevented.

John Paul Lederach
September 25, 2011

Preface

Minority Report, an award-winning science fiction film released in 2001, portrays the work of John Anderson, an officer in the "PreCrime" unit of the Washington, DC, police department dedicated to arresting murderers *before* they kill people. PreCrime relies on three psychics for its predictions. As a result police officers save lives with stunning accuracy.

This level of predictive capacity has not been achieved in real life, including in providing an early warning of violence among groups of different identities, based on, for instance, ethnicity, religion, tribe, or gang membership. But there are signs that efforts to prevent violence (what some analysts call *conflict early warning and early response*—to highlight the importance of timely detection of escalating tensions which trigger a rapid response) is becoming increasingly effective, not because of psychics but due to the use of *events data* (coded information about events of cooperation and conflict), *pattern recognition* (mathematically identifying trends in events data), digital maps, radio communications, and *crowdsourcing* (a process whereby people voluntarily send information via text messaging, video feeds, and the like, which are pulled together to provide an overall picture of an unfolding situation). While international observers have been frustrated by the lack of state response to warnings of intergroup violence, there is new hope that community-level interventions based on early warnings can save lives. Private citizens— often with the support of local governments—are taking violence prevention into their own hands, rather than waiting on national governments, foreign governments, or international organizations to respond when conflict escalates.

While there have been substantial advances in techniques employed to detect incipient violence, much less is known about what to do about it. Students of conflict resolution, peace studies, international relations, political science, political geography, social work, and global health, among other fields, often walk away from courses with greater understanding of why violence happens. It is rare, however, that they learn what to do about it. It is even more infrequent that they acquire knowledge about techniques of detection and options for intervention that can have a direct impact on saving lives.

This book seeks to help fill that gap, to address at least part of the hunger many feel to know "What can I do about it?" to complement the question "Why does it happen?" in the all-too-amorphous quest for peace. It also assesses critically the value added by various technologies used to prevent violence. It covers early "low tech" approaches to violence prevention (namely, *community organizing*—that is, facilitating the formation of civil society entities for purposes of social action and development) through evolution into more "high tech" approaches (involving the use of events data, pattern recognition, digital mapping, and social media).

This book has two undercurrents flowing through the pages. One relates to the *level of leaders* engaged in violence prevention. Initial efforts—promoted by international relations scholars and diplomats—at early warning and early response were conceived and implemented at a macrolevel. The idea that outsiders could monitor tensions within or between other countries and then marshal an effective, timely intervention has, generally speaking, had a disappointing record. There are a host of reasons why this is so. First, intervening to prevent violence can be a messy. Countries who commit peacekeepers can fall into a quagmire involving casualties and embarrassment, on the one hand (as was the case in famine-stricken Somalia in the 1993, when the world watched in horror as dead U.S. peacekeeping troops were dragged through the streets of Mogadishu after getting caught in interclan fighting), and unclear mandates and embarrassment, on the other (as in Rwanda in 1994, where peacekeepers of the United Nations Assistance Mission for Rwanda did not have authorization to intervene to prevent a genocide unfolding around them, which cost the lives of some 800,000 people). Second, committing peacekeepers quickly when there has been an early warning requires expedited political decision making and associated swiftness in bureaucratic processes. Such speed in making decisions and moving bureaucracies is unlikely when there is considerable risk in responding early. Executive leaders (such as presidents and prime ministers) and their advisers must act with little visibility as to likely outcomes. And bureaucracies must move quickly, something they are not prone to do even when executive mandates have been issued (Allison and Zelikow 1999). Third, international responses are often enkindled, at least within modern democracies, by saturation in the media of human tragedy. Constituencies are moved by images of people suffering from famine, floods, earthquakes, or brutality. Conflict early warning, by definition, is aimed at preventing tragic human conditions from happening. Hence, there is typically no constituency speaking out in favor of early response.

It is understandable, then, that conflict early warning and early response involving top-level leaders, usually international outsiders, has had a disappointing record. In contrast, approaches supporting people at a local level to intervene to prevent violence in their own communities are being noticed and acknowledged. Training people living in communities suffering from considerable intergroup tension, warning them that violence is likely, and then facilitating initiatives to prevent it or to flee from an attack can and has worked.

Indeed, a common view among some scholars and practitioners of peace and conflict is that the focus of conflict early warning and early response should be on the "grassroots." The thinking is that the international outsider approach has not worked, so the focus should be on the local level where it has worked. But as with undercurrents in a river, this one is encountering a backflow. This book explains that the change is not from "top down" to "grassroots." It is from "grassroots" to "middle-out" (linking the "grassroots" with moderate leaders at middle and top levels of political, civic, and religious leadership).[1]

Governments and other international outsiders continue to be interested in forcasting when nation-states will evolve into massive violence. The undercurrent related to level of leaders should not be misunderstood as moving away from an interest among international outsiders. Quite the contrary—there is intense interest and increasing sophistication of methods to detect "state failure" (Goldstone, 2008, 2009; Goldstone et al. 2010). But the undercurrent—the thrust—is moving toward multiple levels of leadership, a trend that has been unfolding for years. This book covers that trend.

The other undercurrent relates to *technology*. Engaging people at a local level through community organizing has been an effective approach well before the use of text messages and social media. There is a sense in which newer high tech approaches—involving lots of people through crowdsourcing, encouraging observers to submit information about tension and conflict and cooperation via text message, social media feeds, and Internet form submissions—can be a substitute for organizing. While these technologies can be extremely helpful, we are learning that they are most effective if combined with sophisticated strategy and effective organizing.

The technology undercurrent mirrors the level-of-leadership undercurrent in that they both encounter backflows. Older approaches to conflict early warning and early response are complemented, not replaced, by newer ones. It is one thing to create a digital mapping platform that depicts information from text messages, social media communications, and Internet form submissions about events during, say, post-election violence. It is quite another to train people in how to intervene to prevent violence and to have high enough confidence in observed patterns to issue warnings that violence is likely. Making quick decisions about warnings in volatile situations is enhanced by virtue of the trust bestowed upon trainees (usually unpaid) who submit events data. They are part of a "trust network." It is reasonable to consider the information they send as valid.

Conversely, sending warnings to untrained people can cause panic and, even worse, can exacerbate tensions that could inadvertently result in causing rather than preventing violence. It is preferable that warnings go to those who know what to do with them and can handle them with discretion. A crowdsourcing approach that engages people in a common events-data reporting initiative, while an important technological advance for the purposes of violence prevention, is improved substantially by having a trained cadre of "people on the ground"—people who are organized. This is important both for getting reliable information (sourcing), and for discrete dissemination of warnings (feeding). That is to say, *bounded crowdsourcing* (sourcing through a network of trusted volunteers or staff members, or both, as well as the crowd), *bounded crowd feeding* (sending information back to those volunteers, staff members, and the crowd, especially those close to the incipient violence who sent the initial data), or *restricted feeding* (that is, sending warnings only to select individuals, mainly trained volunteers and staff members, along with trusted officials, primarily due to concerns that warnings can cause panic and hysteria) is superior to *crowdsourcing* and *crowd feeding*. Training people still matters. Organizing still matters. Strategy still matters.

Make no mistake: crowds who participate spontaneously in sending messages relating to recent developments are an asset. The information they send is valuable, especially when rapid verification processes are employed. Involving lots of people through crowdsourcing matters. It is a quantum leap in participation and potential engagement. Visualization of data on digital maps, populated with information from text messages, social media feeds, and Internet form submissions, potentially analyzed with pattern recognition mathematics, can be helpful. Such compilation of data offers summary pictures of what are otherwise scattered, incoherent flows of information.

In the case of violence prevention, however, the adage "if you build it, they will come"—when applied to building a website that depicts events data on a digital map and urging people to send information to populate the map to be analyzed for patterns—is better worded "if you build it and train some of them, they will lead."[2] We are learning that it is organizing and strategy *with* new technologies that is most effective, tailored for specific types of volatile situations. *Strategic nonviolence*, which often involves bringing about political change by increasing, rather than decreasing, societal tensions, but with little or no bloodshed, can certainly benefit from new technologies (like crowdsourcing) but also requires considerable organizing, strategy, and leadership. In Tunisia and Egypt in late 2010 and early 2011 (where there was violence during the overthrow of the Ben Ali and Hosni Mubarak governments but it did not escalate to massive bloodshed as in Libya), both undercurrents were evident—the ebb and flow of technology and organizing, and the involvement of multiple levels of political, civic, and religious leadership, including outsiders. The youth organizers in Egypt, for instance, received guidance from leaders of youth movements in Tunisia and Serbia, as well as a group of young Egyptians based in Qatar at the Academy of Change. And international outsider governments—even those that had supported the dictators being thrown out for decades, like the United States—used diplomatic influence to facilitate regime change.[3]

The young people in Tunisia and Egypt who were at the vanguard of the "Arab Spring" (pro-democracy and anti-dictatorship movements in the Middle East that became widely visible in late 2010, mirroring "European Spring" liberation movements in Central and Eastern Europe following the downfall of the Soviet Union) did not simply step into a civic vacuum using social media to overthrow dictators. They were organized. They had a sophisticated strategy, developed over a period of years. They seized an opportunity for political transformation fraught with uncertainty, uniting through social media, stifling violent troublemakers whenever possible, communicating with moderate mid- and top-level leaders, involving those at all levels of society, using grassroots organizing in combination with crowdsourcing techniques.[4] They inspired the world with their bravery and organizational prowess in the limelight of duel undercurrents, facing a future bound to be filled with setbacks and frustrating challenges. The undercurrents of evolving technology and the enhanced involvement of leaders at all levels in preventing violence together take us to new places.

This book explains how these undercurrents contribute to exciting possibilities for the future in our efforts to prevent the horror of violence and to promote the transformation of conflict to bring forth more just and equitable societies.

Why does this matter? Imagine you're living in a neighborhood, town, or village filled predominantly with people of your own ethnic and cultural background. Suddenly you and your neighbors are attacked by people with different ethnic and cultural traits who live outside your area. Then you hear a rumor that you are about to be attacked again. Would you pick up your essential belongings and leave? Would you join in with your neighbors to prepare to defend yourself and your loved ones? Would you join a vigilante group to launch a surprise attack on those who allegedly plan to attack you? Would you prepare food and drink for the attackers with a strategy of countering hatred with love?

Or, what would you do if you were appointed to serve on a peace committee in your neighborhood, and you and other committee members heard of an imminent attack? Would you call the police? What if the police are notoriously "partisan" in looking the other way when certain groups attack your neighborhood? Would you call the army? Or how about people you know—friends, colleagues, or neighbors, perhaps—who are reasonable and are of the same ethnic, religious, cultural, or racial identity of the group that is allegedly planning to attack? Would you ask those in uniform whom you trust to intervene? Or would you call prominent political, religious, or civic leaders asking for help?

What if you worked for a non-governmental organization (NGO) and your mission is to prevent violence? How would you respond when a rumor of a possible attack is reported to you? What if you also heard that a paramilitary group—one with a reputation of committing atrocities—is training in the area, and the day before there had been a massacre in a nearby neighborhood, town, or village?

Or perhaps you are a government official or diplomat, concerned about violence throughout the country. Your office is flooded with rumors every day, through stories in newspapers and on the Internet, radio, and television? You also hear of instances where moderate leaders are standing up to extremists who are inciting violence. You wonder if the situation is getting better or worse. Are there places in the country or region that are especially troublesome? Are there patterns to what would otherwise appear to be scattered, random violence?

How do you sift through mountains of information and make sense of it all? How do you build a communications link between the members of peace committees on the village or neighborhood level and the top-level decision makers?

Assume you studied math, statistics, political geography, epidemiology, or computer science in college and you are deeply distressed by violence in your community. Would you be interested in using your computer savvy to track data on deadly violence and cooperation? Would you be willing to work side by side with peacebuilding specialists who are prepared to intervene in situations of imminent violence, accompanying them into situations in which, if efforts to mediate fail, you could be killed?

Or, assume you studied political science, conflict resolution, or peace studies in college and you know how outbreaks of violence can render months, if not years, of work virtually useless by resurrecting memories of past animosities, fears, and insecurities. Would you focus exclusively on strategic issues that address the root causes of the conflict? Or would you also be prepared to "put out fires" that either result from a random incident or are facilitated deliberately by *spoilers* (those who disrupt peace processes out of boredom, thrill seeking, feeling left out, or who otherwise benefit from violence) so that your hard work on the larger issues has the potential to endure?

Imagine that you have lived under the weight of a dictatorship for decades, that you have seen members of your family humiliated, and that you and some of your friends have been reading about nonviolent methods used by Mahatma Gandhi and Martin Luther King Jr. that have been elaborated upon by Gene Sharp, whose books and other publications you pulled off the website of an organization he founded, The Albert Einstein Institution.[5] You recognize that a nonviolent approach to regime change will require incredible discipline; if people become violent toward the police and the military, your strategy will fail. You realize that it is essential to have an early warning system to issue alerts of potential violence. You recognize that you must develop an organized capacity to intervene quickly in an attempt to prevent violence or to keep it from escalating. You hear about how some ingenious "techies" and journalists in Kenya created Ushahidi (a Swahili word for testimony), an organization that provides free software downloadable from the Internet that helped them keep track of post-election violence.[6] You download the software and read blogs about how to use it. Would you have enough confidence in your friends, in your strategy, in your high-tech tools to step out into the streets in nonviolent protest? Would you think twice about it after getting pelted with rubber bullets? Would you reconsider as you watch your friends (as the young people in Egypt's Arab Spring did) prepare to "go into nonviolent battle," placing flattened pieces of plastic bottles under their clothes as "body armor," packing rags (along with lemon juice and vinegar) for use when surrounded by tear gas, wearing bicycle and motorcycle helmets, and carrying trashcan lids as shields? Would you be able to control your temper when you see people being beaten with batons? Would you have enough spiritual centeredness to prevent the poison of hatred from entering your heart?

Acknowledgments

Since 1993, I have had the privilege to witness the work of courageous and committed people standing for justice and promoting peace in India, Pakistan, Jerusalem, West Bank, Israel, Gaza Strip, Guinea, Sierra Leone, Rwanda, Somalia, Bosnia, Macedonia, Serbia, Kosovo, Croatia, Montenegro, Nepal, Sri Lanka, Afghanistan, Iraq, and Ethiopia. It is hazardous to try to list those who have taught me in these fascinating and sometimes troubled places, but here is an attempt. I apologize in advance for inadvertent omissions.

First and foremost, I am grateful to the staff members of the Foundation for Co-Existence (FCE) in Sri Lanka—especially Kumar Rupesinghe, Shevanthi Jayasuriya, Dinidu Endaragalle, Madhawa Palihapitiya, Priyan Seneviratne, Hemantha Bandara, Janaka Bandara, H. D. K. Chathurani, R. E. S. Croos, Damindra De Silva, Nishantha Pushpakumara, Heshana Ranasinghe Sheranga, Roshan Rangajeewa, Wimali Tissera, Abhirami Vanniyakulam, and Menaha Velauthama. Thank you for your generous hospitality extended to both my wife Sue and me and, especially, for your inspirational work that convinced me of the potential of human agency in preventing violent conflict.

I also thank the staff of The Asia Foundation for its assistance and support during repeated visits to Colombo and field research in the Eastern Province. In particular, I am grateful for the help of Nilan Fernando and his dedicated staff members in Sri Lanka, and Kim Ninh, The Asia Foundation's representative in Vietnam. I also appreciate the support of Erik Jensen, Matt Nelson, Debra Ladner, Bill Cole, Gordon Hein, Nick Langston, and George Varughese who were instrumental in implementing a project on democratic governance and conflict management, with funding from the Hewlett Foundation, which provided a basis for my work in conflict early warning and early response. I benefited greatly, as well, from insights offered by staff members of swisspeace, especially from Dominic Senn. Finally, I am grateful to the British High Commission, especially Steve

Ainsworth, for insightful feedback on numerous reports and for the support of Patricia Lawrence and Timmo Gaasbeek in conducting an impact assessment of FCE's work.

I am grateful to the American Refugee Committee (ARC), especially Daniel Wordsworth, Colleen Striegel, and Eric James, for the opportunity to assist with ARC's emergency response initiatives following the devastating earthquake in Haiti in 2010. It was while working in Haiti that the value of crisis mapping became increasingly clear to me.

I am indebted to Clark McCauley for his guidance regarding the social psychological literature on violent conflict. It has been a pleasure working with him on articles and a book chapter, while also enjoying his family's hospitality in Pennsylvania. I am also grateful to the Andrew Mellon Foundation for being instrumental in our collaboration.

Thanks go to Robert Cheetham of Azavia, Dennis Culhane of the University of Pennsylvania's Cartographic Modeling Laboratory, and Mike Urciuoli of the Philadelphia Police Department for their insights about early warning in general and the Crime Spike Detector ("Hunchlab") in particular. I am also grateful to Donna Ruane, my extremely capable colleague at Haverford College, for her support and enthusiasm.

Thanks to Emmanuel Bombande, executive director of West Africa Network for Peacebuilding (WANEP), for his help in understanding the work his organization has been doing with the Economic Community of West African States (ECOWAS); and to Ameth Diouf, regional technical adviser for West Africa of Catholic Relief Services (CRS), for making the introduction. I am grateful to Frank Carlin, Pat Johns, Ken Hackett, Bill Canny, David Holdridge, Phil Oldham, Geri Sicola, Tom Bamat, Jean Baptiste Talla, Mary Hodem, Chris Tucker, Rich Balmadier, Gary Shapiro, and others at CRS for deepening my understanding of how non-governmental organizations (NGOs) can be helpful in preventing violence, including the incredible experience of participating in the Working Group on Reconciliation of Caritas Internationalis.

I am grateful to Shavkat Kasymov, Robert Perera, and Senait Tesfamichael for research assistance, and to Cathy Laake, Dillon Gonderman, Kristi Haas, Andrew Masak, Katie Bilek, Alicia Quiros, Caitlan Foster, and Wing Wong for administrative support and assistance in citation retrieval and graphical design. I am appreciative of communications I have had with Joe Bond and Doug Bond about the Conflict Early Warning and Response (CEWARN) project in particular and early warning in general. In addition, I am grateful for the hospitality of Ambassador Abdelrahim Ahmad Khalil, director, and Betty Abebe, information technology and data project manager, of CEWARN during a meeting in their office in Addis Ababa.

I am appreciative of the useful feedback provided on draft chapters for this book from Casey Barrs, Joe Bond, Ian Delarosa, Damindra De Silva, Luise Druke, Amy Erikson, Patrick Meier, Erik Melander, Desiree Nilsson, Paul van Tongeren, and Peter Wallensteen.

I am grateful to colleagues at the Kroc Institute for International Peace Studies—Scott Appleby, Hal Culbertson, Christian Davenport, Jerry Powers, and Emad El-Din Shahin—for their support and encouragement while I wrote this book. I am grateful, as well, for the support provided by Malinda Yarnell, Jenny Miller, Kimarie Merz-Bogold, Kathy Taylor, Dave Severson, Edwin Michael, and Juan Carlos Guzman of the Eck Institute for Global Health, and Michael Clark, statistical consultant of the Center for Social Research at the University of Notre Dame.

I would be remiss not to mention the two founders of the International Network of Crisis Mappers—Patrick Meier and Jennifer Ziemke. They have generously shared insights via email and at the International Network of Crisis Mappers conference at John Carroll University in 2009 and at the humanitarian summit hosted by the Harvard Humanitarian Initiative in 2011.

I acknowledge with appreciation the Faculty Research Grant provided by the University of Notre Dame from 2008 to 2009, and the support of Frank Carlin and Pat Johns of Catholic Relief Services who allowed me to participate in a fellowship with the W. K. Kellogg Foundation that lasted from 1993 to 1997. It supported work with Mary B. Anderson, Fr. Cedric Prakash, Pushpa Iyer, Marshall Wallace, and others with the "Do No Harm" and "Reflecting on Peace Practice" projects of CDA Collaborative Learning Projects. I am also grateful to the Hanns Seidel Foundation for supporting my participation in a workshop in Karachi, Pakistan, in March 2009, led by Moonis Ahmar, chair of the Department of International Relations, University of Karachi.

I am indebted to Marguerite Avery, Kathy Caruso, Julia Collins, and Katie Persons at the MIT Press, for their encouragement, excellent guidance, and editing. It has been a pleasure working with them, and this book is much better now than it would have been without their support.

Finally, I am grateful especially to my dear wife, Sue, for providing substantial support in editing drafts, asking great questions, and patiently allowing me to spend weekends working on this book despite the lure of the beautiful St. Joseph River behind our house with two kayaks sitting in our garage. I also appreciate the fascinating insights of my son, Nick, about decision making with imperfect information, and the helpfully blunt comments about this book from my daughter, Jessica, who instructed: "Try to use stories to illustrate your points, Dad. Don't be boring!"

Introduction

Social media, which relates to sharing of information, experiences, and perspectives throughout community-oriented Web sites, is becoming increasingly significant in our online world. Thanks to social media, the geographic walls that divide individuals are crumbling, and new online communities are emerging and growing. Some examples of social media include blogs, forums, message boards, picture- and video-sharing sites, user-generated sites, wikis, and podcasts.[1]
—Tamar Weinberg, *The New Community Rules*

We are living at a time of dramatic shifts in information and communication technologies (ICTs) that are transforming how we view and engage in the world. Cell phone cameras and social media (such as Facebook and Twitter) turn individuals into on-the-spot reporters, sidestepping what used to be the exclusive purview of journalists and media syndicates. The general public is now capable of taking pictures and videos of violence, sending them to friends, who then distribute these disturbing images onward, providing mass-media-type coverage in seconds. This "citizen journalism" makes us wonder what these new technologies will mean if used in places suffering from civil strife, including in countries where governments are undemocratic, civil liberties are constrained, and injustices spawn civic entrepreneurship for violent or nonviolent change, as they did in Iran in 2009, Tunisia in 2010, and Egypt in early 2011. As aptly put by Philip Howard, director of the World Information Access project:

Technology alone does not cause political change—it did not in Iran's [2009 post-election] case. But it does provide new capacities and imposes new constraints on political actors. New information technologies do not topple dictators; they are used to catch dictators off-guard. . . . The world has seen interest in change from within Iran, and this may prove to be the most destabilizing outcome of the protests. The regime's brutalities streamed around the globe. The world saw the dissent; the regime knows the world saw the dissent.[2] (2011, 12)

Citizen reporting is not, however, simply exposing oppression by governments. It is also broadcasting potential or actual violence between groups of different identities (based on ethnicity, religion, tribe, or gang membership, for instance) at a local level. These images, too, are being beamed increasingly onto our television, computer, notebook, e-ink readers, and mobile phone screens.

ICTs are impacting not just the coverage of violent events, injustice, and oppression, but also our reaction to those events, injustices, and acts of oppression. There is a link between being better informed and taking action, but there is not an automaticity of stimulus-response. The range of reactions varies from local to international levels, from the impulsive to the strategic, from passivity to full engagement.

It would be a mistake to contend that social media is the central force behind pro-democracy, anti-dictatorship movements of the Arab Spring, but there is no question that ICTs added capacities to these movements that allowed protesters to communicate and coordinate in ways that were not possible before. The importance of preventing violence from escalating into massive proportions during strategic nonviolent initiatives highlights the importance of conflict early warning and early response methods and technologies. It is critical for those seeking to overthrow dictators and herald in democracy to keep a lid on troublemakers who have the potential to alienate and anger the police and military. Using violence to bring about political change reduces a movement's public support. Attacking the police and military has the duel problematic effect of making crackdowns by those in uniform palatable and eroding whatever moral high-ground the nonviolent movement had hitherto enjoyed. Preventing troublemakers from leading people down a path of using violence is critical to the success of strategic nonviolent movements and early warning and early response is integral to that prevention.

Origins and Four Generations

As we consider early warning and early response methods and technologies, it is helpful to know their origins. As explained by Anna Matveeva, there are

two main sources—disaster preparedness, where the systematic collection of information was expected to shed light on the causes of natural calamities, and the gathering of military intelligence. In the 1950s a connection was made between the efforts to predict environmental disasters, such as drought and famine, and attempts to foresee crises arising out of political causes. The period from the 1960s to the

1970s was characterized by a firm belief in the value of information technology and faith in the wonders of statistical analysis. Granted large budgets by the governments, projects were constructed which used event data-coding and sought to build models for understanding political behavior. These started to fall out of favor in the mid-1980s. (2006, 9)

The field of conflict early warning and early response has evolved through four generations of approaches. David Nyheim's report to the Organisation for Economic Co-operation and Development (OECD) defines three of them (2008). The first generation of systems was initiated in the late 1990s (and some of these systems are still operating today). These early systems were designed to inform outsiders about potential conflict in other countries. An example is the International Crisis Group's approach before it established regional offices and the European Commission's use of a conflict indicators model.[3]

The second generation, which started in early 2000, was also developed and intended for use by outsiders. These systems, however, incorporate information gathering from field monitors. Qualitative insights are combined with a quantification of events to produce reports. These reports are sent to top-level decision makers so they can take action before an anticipated crisis. Examples include the FAST (a German acronym for "Early Analysis of Tensions and Fact-finding") system of swisspeace (which is no longer in existence due to funding constraints), the Network for Ethnic Monitoring and Early Warning (EAWARN) based in Moscow, and the current approach used by the International Crisis Group (ICG).[4]

Third-generation systems were developed in 2003 and 2004. They are based in conflict zones where field monitors, in addition to collecting information to enter into an events database, get actively involved in violence mitigation and *conflict transformation* (a process that does not necessarily bring resolution of conflict but that at least develops approaches for dealing with it constructively, assuming that tension is part of human relations and that conflict can be channeled in positive directions). Examples provided by Nyheim include the Program on Human Security and Co-Existence of the Foundation for Co-Existence (FCE) in Sri Lanka (described in detail later in this book); the Conflict Early Warning and Response Mechanism (CEWARN) of the Inter-Governmental Authority on Development (IGAD) focused on pastoralist-agriculturalist and pastoralist-pastoralist violence in the Horn of Africa based in Addis Ababa, Ethiopia (also described in detail later in this book); and the Forum on Early Warning and Early Response-Eurasia (2008, 19).

Since Nyheim wrote his report, a fourth generation of early warning and early response has been defined by Patrick Meier. These systems, according to Meier, like those of the third generation, "are also based in conflict areas . . . but there are not pre-designated 'field monitors.'. . . they draw on crowdsourcing [which involves receiving messages from volunteers close to an event] for both early warning and early response [and] on open source, freely available software." Quoting the Third International Conference on Early Warning, he called fourth-generation systems people centered, meaning to "empower individuals and communities threatened by hazards to act in sufficient time and in an appropriate manner so as to reduce the possibility of personal injury, loss of life, damage to property and the environment, and loss of livelihoods."[5] It is important to keep in mind, however, that these systems are mainly being applied to crisis management and have had limited use for early warning and early response purposes.

In my view, it is best to refer to the first two types of early warning and early response approaches as "macrosystems" and the third and fourth as "microsystems." This clumsily mirrors the categorizations in macroeconomics and microeconomics. It is useful nonetheless in clarifying that microsystems focus substantially on information flow and warnings at a local level in addition to information flow and warnings with mid- and top-level leaders. That is, the flow is both horizontal (local to local) and vertical (local to middle and top levels), whereas with the macrosystems of the first two generations the information is almost exclusively vertical, aimed primarily at alerting top-level government officials within the country of tension (at least the moderate leaders of those governments), external governments, and transnational organizations. By using these terms, we can forgo an inadvertent and undesirable inference that first- and second-generation systems lack utility when they are appropriate and needed for specific circumstances.[6]

The Warning-Response Problem

A major critique of conflict early warning and early response systems, broadly speaking, is that while producing warnings is relatively easy, marshaling an early response is not. This criticism has been leveled mainly at macro systems. George and Holl call this the *warning-response problem* (1997; see also Lund 1998).

There are, in fact, indications that the warning-response problem is being solved, at least partially, with macro systems. Transnational organizations and governments are using satellite imagery to track massive atrocities and bring international pressure to bear to end them. United States Agency for International Development (USAID), with support from Google Earth and the United States Holocaust Memorial Museum, tracks the destructive behavior of the Janjaweed militia in Darfur, Sudan. Similarly, in southern Sudan, the Satellite Sentinel Project, founded by actor George Clooney and Enough Project cofounder John Prendergast, combines field reports and satellite imagery with Google's Map Maker for an early warning system of mass atrocities. This project is the result of collaboration between the United Nations Institute for Training and Research (UNITAR) Operational Satellite Applications Programme, Not On Our Watch, the Enough Project, Google, DigitalGlobe, the Harvard Humanitarian Initiative, and Trellon, LLC.[7]

While this technology works well for observing patterns of major violence, the daily update of data by satellites in non-geosynchronous orbit (which remain fixed in space without rotating with the earth) are less useful for early warning. Hence, outsiders are starting to use unmanned aerial vehicles for more regular surveillance. Small planes equipped with cameras have been used in Darfur, along with satellites (Kreps 2010; Leaning 2010). Similarly, in an initiative called Voix des Kivus, implemented by Columbia University with support from USAID, text messages from remote villages are being used to track violence in the Democratic Republic of Congo, and the information is being used for both local violence prevention and top-level initiatives aimed at putting pressure on belligerent parties.[8]

Unmanned aerial vehicles can fly over an area, secure high-resolution images, and can be redirected in ways that satellites cannot be. Current versions of unmanned aerial vehicles are obtrusive, but very small versions, using both stealth technologies and nanotechnologies, are being developed.[9] These technologies will allow outsiders to see what is happening without creating diplomatic damage with an allegedly genocidal government or "warlord" (Kreps 2010).

It is clear, however, that the warning-response problem that plagues macrolevel systems that involve international outsiders—namely, government officials and staff members of UN organizations—is, at least in part, a function of political and bureaucratic constraints. Carment and Garner (1998) and Cockell (2003) insightfully point out how a major priority in developing macrolevel early warning and response systems must be to apply decision-making theory to program design, so that decisions can be streamlined. Hoffman and Hoffman contend that INGOs can intervene successfully to prevent macrolevel violence, working closely with host country governments (2006).[10]

Political and bureaucratic constraints that typically constrain foreign governments and transnational organizations, in contrast, do not substantially affect early warning and early response initiatives at a local level. Non-governmental organizations (NGOs) and community-based organizations (CBOs) are responding to warnings and preventing violence within their communities.

Microfinance Compared to Microconflict Innovation

So the field of violence prevention is in flux. There is a sense in which those dedicated to building peace are making advances locally that can have far more success than those focused on violence at a macrolevel (such as during a separatist revolt or other widespread insurgency). This enthusiasm is not unlike the fervor and passion surrounding the methodological revolution integral to microfinance following the path-breaking work of the Grameen Bank in Bangladesh. Mohammad Yunus, the bank's founder, refused to accept the classical view in economics that poor people are bad credit risks. He reasoned that their social capital—their sense of solidarity— was comparable to financial equity in securing loans. Large banks had proven to be ineffectual in lending to poor people, so he started his own bank, working with local communities of solidarity groups, with small amounts of money, yielding remarkably robust repayment rates.[11] Even though Yunus and others have been criticized for the interest percentages involved in sustainable microlending, there is little doubt that he ignited a financial revolution that put money into the hands of poor people.

In the case of violence prevention, innovators—like Fr. Cedric Prakash and Pushpa Iyer in India; Gary Slutkin and Candice Kane in the United States; Kumar Rupesinghe, Dinidu Endaragalle, and Madhawa Palihapitiya in Sri Lanka; Ory Okolloh, Erik Hersman, David Kobia, and Juliana Rotich in Kenya; and Abdelrahim Ahmad Khalil and Betty Abebe in Ethiopia— have taken early warning and early response to the local level. They challenged the widely held view that outsiders—whether they be officers of transnational organizations, foreign governments, or national governments—need to intervene to prevent violence. They have bypassed these large organizations with their associated political constraints and bureaucratic inertia just like Yunus bypassed large banks. They have gone directly to local communities. They have shown that the warning-response problem at a local level can be solved and, in so doing, have prevented violence successfully.

Strategic Peacebuilding

Identifying incipient violence and intervening to counteract it have been some ofthe most intractable challenges of strategic peacebuilding. *Strategic peacebuilding* is defined by Lederach and Appleby as

an approach to reducing violence, resolving conflict and building peace that is marked by a heightened awareness of and skillful adaptation to the complex and shifting material, geopolitical, economic, and cultural realities of our increasingly globalized and interdependent world. Accordingly, peacebuilding that is strategic draws intentionally and shrewdly on the overlapping and imperfectly coordinated presences, activities and resources of various international, transnational, national, regional, and local institutions, agencies and movements that influence the causes, expressions, and outcome of conflict. Strategic peacebuilding takes advantage of emerging and established patterns of collaboration and interdependence for the purposes of reducing violence and alleviating the root causes of deadly conflict. (2010, 22)

Readers may wonder why violence prevention at a local level, a largely *tactical* enterprise, is a part of *strategic* peacebuilding. Strategic peacebuilding involves achieving justice so that peace is sustainable over time. In contrast, violence prevention is aimed at counteracting the impact of events that are seized upon by those who perceive that they have something to gain from violence. It is more tactical, but nonetheless integral, to strategic peacebuilding. Think of it this way: in building a city, with huge investments in residential and commercial structures, would you not have a fire department? Figuratively speaking, if extensive effort and investment is made in peace, would it not be prudent to put into place a systematic method for detecting and responding to provocative situations that can lead to violence? Similarly, in developing the organizational infrastructure and strategy to topple a dictator using strategic nonviolence, wouldn't it be important to be prepared to nip violence in the bud, especially given that violent outbursts during macrolevel political transformation are going to happen? What kind of violence suppression infrastructure can you build? How do you move from the notion, the idea, the goal of nonviolence to the apparatus of nonviolence?

Efforts to prevent violence must be a part of any strategic peacebuilding strategy, including those that involve the use of strategic nonviolence, as we have witnessed in the Arab Spring. Violence prevention is critical in getting traction on what is always an uphill challenge when struggling for sustainable peace in divided societies.

Violence unfolds in a cyclical, not linear, way. So it is not surprising, though it is profoundly disappointing, that many countries, after having gone through painstaking negotiations for years, revert back to massive violence. The World Bank estimates that approximately 40 percent of countries which have negotiated a peace agreement experience a recurrence of substantial violence within ten years.[12] Indeed, Collier, Chauvet, and Hegre (2008) find that of the countries which have suffered significant conflict but have negotiated a peace agreement, 42 percent revert back to massive violence. Even 27 percent of these countries which have experienced at least ten years of sustained economic growth continue to experience a resurgence of significant violence.

Strategic peacebuilding must address economic issues and the overall equity of a society, but it is essential also to incorporate ways to address acute eruptions of tensions during fragile transitions from conflict to peace. And just as violence prevention is critical for the success of strategic nonviolence in keeping a lid on troublemakers, it is of similar importance in strategic peacebuilding in a post massive violence phase, in keeping the lid on spoilers. It is instructive to focus on these two broad categories of people who are instrumental in facilitating or perpetuating violence— *troublemakers* and *spoilers*. Keeping in mind the diagram in figure 0.1, troublemakers cause violent conflict to escalate during phase A, whereas spoilers prevent violent conflict from de-escalating in phase C. Violence prevention is strategic in phases A and C. It is tactical in phase B in the sense that mitigating violence at a local level is unlikely to have a significant impact on the wider, massive violence in which the nation-state or region is engulfed. During phase A, violence prevention efforts using early warning and early response must focus on "troublemakers thirsty for blood" and their followers.[13] The troublemakers typically promote the use of violence to overcome challenges of injustice, to achieve domination, or to gain power, prestige, or wealth in some other way. Some of them, at least initially, have noble goals but embrace nefarious means. Velupillai Prabhakaran, for instance, the infamous leader of the Liberation Tigers of Tamil Eelam (LTTE), was familiar with the idea of using nonviolence for social change because his parents admired Mahatma Gandhi and his nonviolent doctrine of *ahimsa*. But he abandoned any interest in it in September 1987 after seeing a fellow Tamil, Thileepan, fast to death without causing the predominantly Sinhalese-controlled government of Sri Lanka to heed to demands for independence for the Tamil minority. Thereafter, Prabhakaran committed himself to "a path of unremitting carnage" until his death on May 18, 2009 (The Economist 2009, 92). What a tragedy that Prabhakaran did not explore other nonviolent methods to use in standing up for the rights of minorities in Sri Lanka. Who knows whether they would have worked; much would have depended upon how they were used, the timing of their use, the exercise of leadership, and perhaps the influence of international outsiders. But it is clear that the path he chose, embracing violence, was an abysmal, tragic failure. The goal of a separate Tamil state was not met, and the disastrous struggle from 1983 to 2009 to create one resulted in more than 70,000 dead and hundreds of thousands internally displaced, not to mention the legions of torture victims and decades of anemic economic activity despite Sri Lanka having been seen internationally as one of the "growth with equity Asian tiger" stars before the carnage began.[14]

Going back to figure 0.1, during phase C, in contrast, early warning and early response are important in curtailing a deterioration of a peace process already set in motion. Post-conflict environments are often rife with governance challenges leaving vacuums of social order and associated

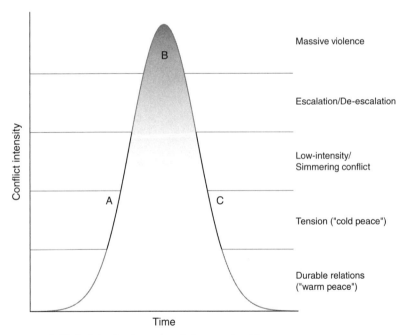

A: Strategic local early response before massive violence
B: Tactical local early response during massive violence
C: Strategic local early response after massive violence

Figure 0.1
Phases of strategic and tactical local violence prevention (adapted from Lund 1996, 38)

lawlessness to those who seize opportunities for looting, raping, extortion, intimidation, and seizing territorial control. Such environments provide spoilers with rich opportunities for profitable "sociopathic entrepreneurship." It is during this phase that spoilers must be dealt with effectively. One approach is to develop reconstruction initiatives that make them have a stake in peace—sometimes called a *spoilers to stakeholders strategy*. Another approach is incarceration of spoilers. And yet another is for those with credibility within the group to discredit spoilers by challenging their rhetoric and appealing to people's sense of wanting to get on with their lives rather than moving back into the disruption, agony and horror of violent conflict. During a period of mass violence, phase B, preventing bloodshed at a local level is important for those whose lives are saved and for those who will suffer the psychological and spiritual wounds incurred in killing other human beings. But the symbolism of the violence at a local level for the wider conflict is much less potent. It is tactical; it can save lives even though doing so is unlikely to have a strategic impact.

Figure 0.2 provides a graphic depiction of violence over time, with each "wave" reflecting the stages presented in figure 0.1, occurring, as pointed out earlier, roughly 42 percent of the time (Collier, Chauvet, and Hegre 2008). Figure 0.2 shows how the wave can keep rising and falling. It is

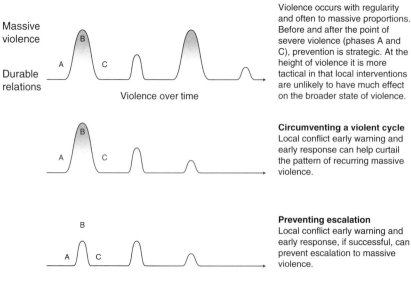

Conflict oscillations
Violence occurs with regularity and often to massive proportions. Before and after the point of severe violence (phases A and C), prevention is strategic. At the height of violence it is more tactical in that local interventions are unlikely to have much effect on the broader state of violence.

Circumventing a violent cycle
Local conflict early warning and early response can help curtail the pattern of recurring massive violence.

Preventing escalation
Local conflict early warning and early response, if successful, can prevent escalation to massive violence.

Figure 0.2
Oscillations of tensions and violence

common to refer to recurring violence as *cyclical*, as I did earlier. In fact, when one plots it over time, it is more precise to describe it as *oscillating*. In the parlance of system dynamics experts, societies plagued by violent conflict encounter competitive feedback loops—one that promotes violence; the other, peace—that seek dominance (see Ford 1999).

But the objective of violence prevention at a local level is to circumvent this cycle, as depicted in the second oscillation in figure 0.2. Or, even better, local conflict early warning and early response, if successful, can prevent a country from escalating into massive violence, as depicted in the third oscillation in figure 0.2.

Preventing violence initially, or preventing it from recurring, is, of course, an imperative from the standpoint of reducing human suffering. Some scholars have also tried to estimate financial benefits to preventing massive violence. According to the Carnegie Commission on Preventing Deadly Violence, for instance, the international community spent roughly $200 billion during the 1990s on various *conflict management* activities— efforts to contain and transform massive violence once it erupts, such as in Bosnia, Cambodia, Somalia, Rwanda, Haiti, the Persian Gulf, and El Salvador. The same report claims that *violence prevention* interventions, when the international community responds quickly to signs of trouble, thereby preventing an escalation to massive violence, could have saved $130 billion (George and Holl 1997; see also DeRouen and Goldfinch 2005, 27).

Much has been written about violence prevention using conflict early warning and early response at macrolevels (see, e.g., Rupesinghe and Kuroda 1992; Davies and Gurr 1998; Goldstone 2008, 2009; and Goldstone et al. 2010). Similarly, there is a growing literature about how to identify violence at a community level, such as in a village or on a street in a section of a city (Adelman 1999, 2006; Ackleson 2006). In comparison, very little has been published about how to respond to early warnings effectively. Recognizing this need, the European Commission funded a five-year research project at the Institute of Development Studies at the University of Sussex in Brighton, UK, called "MICROCON, A Micro Level Analysis of Violent Conflict." In explaining this initiative, MICROCON researchers state: "At a fundamental level, conflict originates from individuals' behavior and their interactions with their surroundings—from its **'micro'** foundations. However, most programs tackling conflict are driven by **regional**, **national** and **international** perspectives. This makes inadequate concession to the role of **individual** and **group** interactions."[15]

Organization of This Book

This book is organized into four main parts. Part I covers general concepts and approaches used in preventing violence. Chapter 1 presents analytical frameworks and theories of violence prevention, highlighting the consensus building that typically occurs before one group attacks another.

Chapter 2, "Reporting and Warning about Deadly Possibilities," describes how ICT is being used increasingly to enhance the effectiveness of interventions designed to prevent violence. It covers how events data are being created and analyzed to track general trends in conflict and cooperation. In this chapter I distinguish early warning from risk analysis, in addition to identifying more sophisticated techniques including those using pattern recognition, automation, web-based digital mapping, hidden Markov models, and crowdsourcing. Chapter 2 explains how attempts are being made to enhance the accuracy of warnings and the speed at which they are issued.

In part II, chapters 3 through 7 cover five cases of violence prevention at a local level. The first case in chapter 3 concerns a low-tech approach used in the slums of Ahmedabad, India. St. Xavier's Social Services Society ("St. Xavier's" for short) forms peace committees and trains the members how to detect early warning signs of violence and how to intervene to prevent it. This case addresses identity-group violence based on ethnicity and religion. It presents evidence of both successes and failures to prevent violence.

Chapter 4 then describes the CeaseFire Project in Chicago and covers its main components, including community mobilization and training of former gang members to intervene before violence erupts. It explains how CeaseFire has developed a social-media virtual reality approach, using Second Life, for training at a distance. It reports the findings of a U.S Department of Justice evaluation of CeaseFire that concludes this program is saving lives.

Chapter 5 describes the Human Security Program of the Foundation for Co-Existence (FCE) in Sri Lanka. This program focuses primarily on preventing violence between Hindu Tamils, on the one hand, and Muslims and Christian Tamils, on the other. Similar to St. Xavier's, FCE works with local civic leaders to form Co-Existence Committees, and trains their members to identify incipient violence, and how to intervene to de-escalate tensions. FCE staff members send reports into the Colombo-based headquarters about events. Those events, as well as others gleaned from media sources, are coded and entered by staff members into a database. This information is used for trend analyses for periodic reports and to inform mid- and top-level leaders so they can bring their authority to bear to help prevent violence, if need be. Pattern recognition is not used and events data analyses are of virtually no use to staff members located where they could potentially intervene to prevent violence. Chapter 5 also reports the results of the members of Co-Existence Committees' and FCE's interventions to mitigate violence when they relied upon their own inductive reasoning without the support of early warning alerts generated using computerized pattern recognition.

Chapter 6 describes how a group of journalists in Nairobi developed an innovative crowdsourcing approach that depicts—on a web-based digital map—aggregations of text messages, information sent over the Internet or through social media, and email reports of post-election violence. This high-tech approach uses various methods to determine data accuracy that are being developed in an open-source, shareware environment. This approach is largely used for crisis management, not violence prevention, but it is a promising technology that may well prove instrumental in both micro- and macroviolence prevention.

Chapter 7 covers the high-tech Conflict Early Warning (CEWARN) Mechanism of the Inter-Governmental Authority on Development (IGAD) that is being used in Djibouti, Ethiopia, and Kenya, Somalia, Sudan, and Uganda. It explains how communication is difficult in remote areas where relations between pastoralist tribes are being monitored. It describes CEWARN's ICT4Peace initiative which uses high-frequency radio, high-gain antenna phones, community radio, and tracking systems to prevent violence related mainly to livestock raiding and rustling. Chapter 7 also points out that the results of CEWARN's approach are preliminary at this stage and highlights evidence that mitigation reduces organized livestock raids, one violence indicator.

Chapter 8 compares and contrasts the approaches in the preceding five cases, assessing their respective strengths and weaknesses, the costs and benefits of each as their levels of technological sophistication evolved over time. It argues that a central feature of a successful violence prevention strategy is the support of local capacity to respond quickly and effectively when tensions begin to escalate. If this local capacity does not exist, then other features, however technologically sophisticated they may be, will be of limited value in preventing bloodshed.

Chapter 9 offers examples of how to intervene effectively to prevent violence. They relate to the analytical approach and theories presented in part I. And chapter 10 concludes part II with this critical point: knowing when violence prevention will probably fail is just as important as knowing when violence is likely to occur. To put it plainly, being prepared to flee is as essential as understanding when and how to intervene.

Part III of this book covers resource allocation considerations and offers recommendations. Chapter 11 explains how critical observers argue that the use of ICT to enhance accuracy and speed of early warnings constitutes a squandering of precious resources, primarily because international aid agencies make their funding contingent upon the use of sophisticated methods; while those seeking to prevent violence do not benefit substantially from the reports those methods generate. Some observers contend that it is tempting to become enamored by technology at the cost of doing the more difficult work on the ground of organizing to build local capacity for effective early intervention.

Chapter 12 looks to the future of how the use of technology might evolve. It, too, relates to resource allocation, and offers ideas about exploiting economies of scale to enhance software. It describes how decision sciences can assist the field of violence prevention in refining our technological approach of when to issue a warning. Chapter 12 also provides guidance on how to minimize costs, and offers recommendations for funders, researchers, and practitioners working on violence prevention.

In the conclusion I offer views about the essential characteristics of people who travel the lonely road of trying to prevent violence in their own communities. Fortunately, this work is not a technology-driven treadmill that goes nowhere; there is solid evidence of success already. But there is much more to be discovered, developed and—most important—implemented.

I Theory and Methodology

The idea of early warning and early response to prevent violence is simple: Find out when violence is about to occur and pursue timely intervention to prevent or evade it. The intervention can involve leaders at a local level (such as a member of a peace committee), middle level (such as a parliamentarian of that part of a country), and top level (such as the head of a government ministry or a diplomat). But when there is only limited time, those nearby who are in the fulcrum of a gathering violent storm are the only ones who can physically intervene to prevent it or convince likely victims to evade it.

In part I, I cover an analytical framework that has advanced our understanding of how local violence unfolds both in the *components* of consensus building to become violent and in the *timing* of that consensus building. This framework has many practical implications for those who work to prevent violence. I compare early warning analyses using dynamic events data, on the one hand, with risk analyses using relatively static indicators, on the other. I explain how events are coded, weighted, and then used for developing reports and warnings.

Part I provides an important foundation for understanding other chapters. Some sections may seem a bit dense and some material is mathematical. But I keep the focus mainly at a conceptual level without complicated formulas.

1 Toward an Applied Theory of Violence Prevention

Many studies from the field of conflict resolution suggest that mediations and other forms of professional third-party interventions have better chances when the conflicting parties are not highly polarized. . . . Many of the conflict mitigation tools have a broad literature, and it has been in depth studied, but usually in relation to both post-conflict situations or advanced stages of conflicts and very rarely to pre-crisis stages lacking at the same time a theoretically conducted and empirically-based analysis of the subject matter.

—Éva Blénesi, "Ethnic Early Warning Systems and Conflict Prevention"

For years social scientists have been developing theories to explain why there is violence between people of different identities. Many of the theories have a mechanical quality. The theory of relative deprivation, for example, posits that it is not so much a lack of resources that makes people become violent. Rather, it is the perception that one group is being favored above another group, and the latter group feels deprived in a relative sense (Merton 1938). Quantitative studies tested hypotheses deduced from such theories, documenting relationships between variables such as inequity between groups and the number of injuries and deaths due to violence (Gurr 1970).

These theories based on physical conditions have since been improved upon by those with a social-psychological bent. Whereas physical phenomena that indicate intergroup inequity are relatively *static*, social-psychological processes of, for instance, consensus building to kill those of another group are *dynamic*. Observations of the types of consensus building that take place are being elucidated with greater precision so that a combination of static and dynamic indicators can be monitored to determine *when* violence is likely to erupt (Bock and McCauley 2003). Those eager to prevent violence are also developing approaches to keep track of rumors and the like, plotting events using digital maps on the Internet and employing geographic information systems (GIS), thereby providing greater precision of information about *where* violence is likely to unfold.

Horowitz's Analytical Framework

Early warning and early response efforts have yielded numerous methodologies about how to identify incipient violence before it erupts. Less well developed are theories that explain or offer insights as to how to effectively produce "actionable" information—that which provides compelling evidence as to when, where, and how to intervene. One noteworthy exception to this is an analytical framework developed by Donald Horowitz of Duke University. The components of this framework, which can be distilled from Horowitz's book *The Deadly Ethnic Riot* (2001), can be divided into three categories: *underlying conditions, time to build a consensus for violence*, and *processes during a lull*. These are listed in table 1.1.

Horowitz bases his analytical framework on an extensive inductive process involving analysis of 150 violent incidents throughout the world. He documents instance upon instance, in country after country, where violence erupted, concluding that "four underlying factors" best explain "the deadly ethnic riot" (2001, 524). The first is "a hostile relationship between two ethnic groups." We know from other social scientists that perceptions about this relationship are subject to manipulation, when the depiction of a history of enmity glosses over periods of cooperation and a sense of tension is exaggerated by those who have something to gain from it (Rudolph and Rudolph 1993; Snyder 2000, 334).

The second factor is "a response to events that engages the emotions of one of those collectivities, a response usually denominated as anger

Table 1.1
Components of Horowitz's framework

Underlying conditions	Hostile relationship
	Precipitating event
Time to build a consensus for violence	Lull, with rumors occurring near the onset of violence
Processes during lull	Acute emotional engagement
	Justification for killing
	Exaggeration of threat posed by the out-group
	Reduction of the risk of violence to the in-group
	A sense that there are no other options

but perhaps more accurately rendered as arousal, rage, outrage, or wrath" (Horowitz 2001, 524). As other authors have noted, certain symbols, particularly religious ones, seem to foster this emotional escalation (Varshney 2001; Bock and Anderson 1999).

The third factor is "a keenly felt sense of justification for killing" (Horowitz 2001, 524). This observation coincides with the theory of moral violation offered by McCauley, which holds that creating a sense of moral indignation is an effective approach for leaders to employ when preparing followers to kill members of another group (2000).

And, fourth, Horowitz identifies a relatively unnoticed factor: "assessment of the reduced risks of violence that facilitates disinhibition" (2001, 524). He documents how potential rioters take great pains to assess the risk of killing, seeking broad social support within their group, ensuring that they have overwhelming numbers, assessing the likely response of police or soldiers, and exploiting the advantage of surprise.

There are two other factors noted in *The Deadly Ethnic Riot* that deserve mention. The first is what Horowitz calls *precipitating events* (2001, 317). These are instances that have considerable symbolic potency, such as when there is a widely held perception that an election has been rigged to the detriment of one group that feels persecuted. The significance of these events can be amplified by those who have something to gain from the politics of division. This is consistent with the findings of Brass (1997), who compares ethnic conflict to a theatrical performance. The first act is often unplanned, but those that follow are orchestrated by an "institutionalized riot system" to make the first act seem more significant than it was. Consider this example: several intoxicated men of one identity group rape a woman of another identity group. The rape was unplanned, but once it becomes known, it is used by those who have something to gain from violence, who propagate rumors that the single rape presages other rapes of the women of one group by the men of the other. This claim leads to another factor that needs to be included on Horowitz's list: exaggeration of threat. Somehow, the precipitating event has to become understood broadly as evidence that the status quo in intergroup relations is going to be disrupted or that otherwise great harm is likely to be forthcoming. The precipitating event is viewed as the tip of the iceberg of a larger threat.

But Horowitz also places the etiology of ethnic riots on a temporal plane, arguing that there is a relatively consistent time lag between the precipitating event and the onset of violence (as depicted in figure 1.1). This he calls the *lull* (2001, 72). It is during this critical interval that

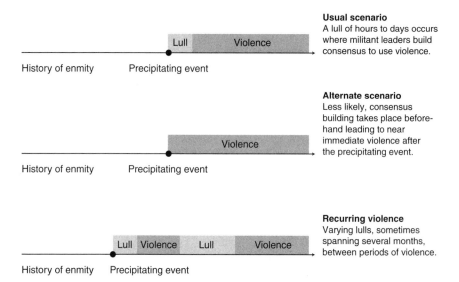

Figure 1.1
Patterns of precipitating events, lulls, and violence

militant leaders facilitate a social-psychological process within an in-group to build consensus to use violence. This is accompanied with scrupulous target identification and risk analysis. Horowitz argues that a lull will be lacking only when this preparatory consensus-building work has been done prior to the precipitating event, causing violence to trigger relatively soon thereafter (as depicted in the second bar in figure 1.1). This latter pattern is less typical than the more frequent one of precipitating event-lull-violence.

An example of violence occurring on the heels of a precipitating event is the riot in Kenya that erupted soon after the December 27, 2007 elections that were perceived widely as rigged. Political leaders affiliated with Kenya's Kikuyu tribe had dominated the country since Kenya's independence in 1963. The reelection of a Kikuyu, President Mwai Kibaki, enraged opposition members of the Orange Democratic Movement, whose leader, Raila Odinga, lost the election, it was believed, due to fraud. Tribal gangs fought violently, killing more than sixty people until the police instated a dusk-to-dawn curfew. It was in this context that journalists combined efforts to track violence using GoogleEarth™, forming Ushahidi, a nonprofit technology company that is covered in greater detail in chapters 2 and 6.

One conclusion that can be drawn from the framework (though it has not been tested) is that all of the components must be in conditions that are conducive to violence for violence to occur. An alternative view is that if one or more of the components can be influenced through various interventions, then violence can be averted by preventing one or more of the components from going into a conducive state. This we will call the *short-circuiting proposition*. Such interventions might include preventing an "engineered" precipitating event from happening; counteracting images of hostile relations; offering constructive mechanisms for channeling anger related to perceived injustices (such as through judicial processes); challenging the rationale behind the justification for killing; discrediting rumors about the severity of the threat; and making it clear that there will be consequences (such as incarceration) for perpetrators of violence (that is, increasing the risk).

When should this "short circuiting" happen? It follows from Horowitz's framework that it must occur during the lull.

One additional component of this framework has been suggested by John Paul Lederach. In his experience, it is common for people to feel "trapped" when they are experiencing acute tension.[1] They often have the sense that there are no other options but violence in response to the tension. I have added this to Horowitz's framework as an additional component in table 1.1.

Time to Respond

On August 14, 2004, I interviewed Brigadier General Sunil Tennekoon in his office in Trincomalee, Sri Lanka. He led a peacekeeping mission during the tenuous ceasefire between the Government of Sri Lanka (GoSL) forces and the Liberation Tigers of Tamil Eelam (LTTE). He said that, in his experience, the Sri Lankan Army had "about five minutes" to intervene with a show of force to prevent bloodshed after it found out that a violent incident was about to occur.

Was this the extent of the lull, or was its beginning discernable earlier? If the lull is of such short duration, or even somewhat longer, only those who are in close proximity to the place of likely violence are able to respond physically. Of course, phone calls can be made and text messages can be sent. But are there times when the physical presence of those who do not work or reside locally is required?

In contrast to the average reported length of warnings of impending violence received by this peacekeeping military commander, Horowitz concludes that a lull, when consensus for violence is being developed, tends to be longer: "Precipitants do not always evoke an immediately violent response. There is often a period, measured in *several hours* or a *few days*, during which the impact of the precipitating event is felt and forces are mobilized for the assault. . . . Occasionally, the period of suspended action may last as long as a *week*" (2001, 89; emphasis added).

In another instance, when referring to the *recurring* nature of ethnic violence, Horowitz observed an even longer period. Commenting on the violence in Assam, India, in 1955, he observed that the lull continued over a period of *months*, involving protests and counterdemonstrations that spiraled into violent conflict (2001, 280; emphasis added).

So we can conclude that Horowitz estimates that instances of ethnic violence usually have a lull of between a few hours and a few days, lasting up to a week (as depicted in the first bar in figure 1.1). In contrast, by his estimations, those instances that recur can have even longer lulls, measured in months (as depicted in the third bar in figure 1.1).

Another researcher of ethnic riots identifies a similar duration of lulls. In his review of Horowitz's book, Kakar noted from his observations in India that lulls tend to last from "12 to 24 hours" (2001, 2097).

While these conclusions are helpful, the proposition that a lull has a relatively predictable length is something that can be tested empirically. And as we assess the efficacy of various approaches to early warning and early response, the length of a lull is a central factor. It relates directly to where emphasis should be placed, and who can reasonably be expected to intervene, to prevent violence. As depicted in table 1.2, using Lederach's (1997) triangle of three intervention levels, if General Tennekoon was right, then we can make an a priori determination that efforts should focus primarily on building and supporting violence prevention capacity at a local level (what Lederach calls "Track III"), with a secondary focus on linkages with a network of key mid- and top-level ("Track II and I") leaders who can be influential with phone calls and text messages by urging their subordinates in closer proximity to take preemptive action. If, however, lulls tend to be longer, and detecting consensus building is relatively straightforward early in a lull, then a much wider range of early response options are plausible. For instance, mobile units of peacekeepers on motorcycles or using helicopters could provide a physical show of force to short circuit consensus building. Alternatively, peace committees can contact those with more influence so that mid- and top-level moderates can prevail upon extremists, persuading them that it is not in their best interests to undertake violence.

Table 1.2
Levels of leadership and response capacity

Level of intervention	Leaders involved	Approach (long lull)	Approach (short lull)
Level I: Top-level	Government officials Diplomats	There is time to order an intervention by police or peacekeepers even if they are positioned far away from the likely conflict.	Little, if any, impact, unless police and peacekeepers are nearby.
Level II: Mid-level	Higher-level religious leaders Civic leaders Intellectuals NGO directors	There is time for mid-level leaders to seek the involvement of top-level leaders.	Not much time to reach out to top-level leaders.
Level III: Local-level	Religious leaders Community-based organization leaders	Focus on local-level leaders to work in partnership with other actors so there is sustainable capacity to counteract the influence of troublemakers and spoilers.	Emphasis on local-level leaders if the lull is short since there is very little time to marshal a multilevel response.

In other parts of this book, I will refer to these three levels of leadership. I will refer to them as local-, mid-, and top-level leaders. Those who are more used to the terminology of Track III, Track II, and Track I leaders should keep in mind that these are the same distinctions about the level of leaders. I use the terms "local-level," "mid-level," and "top-level" for simplicity.

There are, of course, numerous analyses that use these terms when advocating for one approach or another to prevent violence. Is a bottom-up approach best (i.e., from the local to the top level of leadership)? Is a top-down approach optimal? Or, is middle-out the most effective (i.e., approaching mid-level leaders who then advocate for early response with local- and top-level leaders)? In this respect, I find Currion's perspective helpful: "One of the few lessons that has been learnt by organisations working in ICT4Peace is that top-down solutions rarely work. We've learned it so well that we've failed to realize that the reverse is also true: that bottom-up solutions rarely work either. There's a balance to be struck between the emergent activities of individuals and the directed activities of organisations, especially when we're chasing the latest technology down the rabbit hole" (2011, 40).

The Greed-Grievance Nexus

In order to understand how consensus to use violence builds, we need to distinguish between underlying conditions and occurrences during a lull. Underlying conditions are referred to by some scholars as grievances. A grievance is when those of an in-group perceive that they are victims of discrimination, prejudice, and injustice and feel collective frustration and anger toward those of an out-group who are seen to be sustaining these inequitable conditions by their hold on power or are simply seen as benefiting from that inequity. It is natural that groups of people develop grievances—based upon perceptions of being treated differently—against those who are not a part of their group who are seen as perpetrators. This can be because a group feels it is discriminated against (for instance, when its members tend not to get their proportionate share of high-paying governmental jobs), prejudged (such as being accused of crimes without cause), or treated unjustly (when, for example, fair and transparent judicial due process is not accorded to members of their group as it is for others, or when there is bias in the allocation of resources or opportunities).

Grievances lay a foundation upon which conflict can build. But to result in violence, that foundation is usually combined with some sort of political action that fosters discontent and channels that discontent to bring about either violent or nonviolent change. And in some cases those who foment discontent have something to gain from the violence. For instance, during my field research in Ahmedabad, India, from 1994 to 1997, I identified three groups that benefited from slum violence: politicians, real estate developers, and militant groups (which are covered in greater depth in chapter 3).

The combination of calling attention to inequity and injustice (grievances) and fanning the flames of discontent once a precipitating event has occurred or is anticipated is what some social scientists call the "greed-grievance nexus" (see, for instance, Berdal and Malone 2000; Collier and Hoeffler 2004; Korf 2005; Bragg and Shaykhutdinov 2007). While the foundation of inequity and injustice accounts for a grievance among those who feel as if they are treated as second-class citizens, people tend not to take collective action, whether violently or nonviolently, unless they are led to do so by someone who has something to gain (greed) from leading people in that direction.

The word "greed" is clumsy for this purpose. It connotes selfishness and personal gain. So it is a rather inadequate word to account for the motivations of extremist leaders, many of whom stand for justice in ways that are self-sacrificing, incurring considerable personal costs. Sure, a political leader might build support for his or her political party by fanning the flames of ethnoreligious hatred, but not all politicians use public office for personal gain. Many believe in the core objectives of their respective political parties. Similarly, some religious leaders view ethnoreligious domination by those of their faith, culture, or race (or all three) to be a noble, if not sacred, goal. Business and civic leaders, however, might be in pursuit of relatively immediate financial gain through violence, but they may well see their businesses as pivotal for overall economic development, a larger societal good. I write these perspectives about extremist leaders not to glorify their behavior but, rather, to provide a view of their likely ways of justifying their actions to themselves and others.

Iniquity, Honor, Duty, and Dissonance

While the word "greed" can do a disservice in describing the motivations of those leading change in the face of injustice, when it comes to religion there can be a different dimension altogether in their motivations. Religion is what some analysts call an *accelerator* of conflict. As Gurr and Harff explain, events coupled with accelerators often develop a momentum of their own, resulting in rapid escalation of tensions (1998, as cited by Payson Conflict Study Group 2001, 5). McCauley explains that sometimes religious, political, and civic leaders not only argue that people are *justified* in killing those of a different ethnic identity, but also that they have an *obligation* to kill "the others" (2000; see also Stewart 2009, 44–45). The group's honor has been challenged; a moral violation has occurred. "Teaching a lesson" or seeking to eliminate an out-group is not simply a matter of correcting a wrong; it is standing against the desecration of that which the in-group holds sacred. McCauley calls this *iniquity theory*, meaning that when religion is involved, leaders with credibility within a faith community can artfully appeal to a sense of honor and sacred duty, urging followers to become violent when facing an offense to what is deemed sacred. He contrasts this with *equity theory* which, like relative deprivation, holds that perceptions about inequities of land, job opportunities, and the like mechanistically result in tensions that can lead to violence as those who feel slighted seek fairness (Bock and McCauley 2003).

Iniquity theory is useful in understanding the emotional pitch of believers who feel something they hold sacred has been dishonored, a heinous offense. Responding to it can be understood as a sacred duty (hence, the phrase "killing in the name of God").

The destruction of property that occurred in Shantinagar, Pakistan, in 1997 provides a vivid example of how violence unfolds along religious lines when something sacred allegedly has been violated. I was there in the aftermath and was struck by how quickly the rioting unfolded. The violence was related to differences in economic well-being between religious groups as well as intense feelings concerning the selling of liquor.

Liquor is illegal for Muslims in Pakistan, but the government has a tolerant policy toward people of other faiths. Christians can get an official document signed by a member of the clergy attesting to their faith that allows them to purchase a set quantity of liquor per month. (My friends in Pakistan and I affectionately called these documents "infidel permits"!) Since some Muslims choose not to follow the Qur'an-ic prohibition against alcohol consumption, they provide a lucrative market for covert liquor sales by a small percentage of Christians who get involved in "bootlegging."

Since Christians, a small minority, suffer from discrimination in Pakistan, relegated to jobs as street sweepers and trash gatherers, they tend to be among the poorest of the poor. There are exceptions, however, to this general rule, including the Christian village of Shantinagar in the Punjab Province. These roughly ten thousand villagers were living at a socio-economic level above the surrounding Muslim villages, mainly because Christian schools provided higher-quality education and relatives with well-paying jobs overseas sent money back home.

On February 6, 1997, Shantinagar and a tiny Christian extension of that city, Tiba Colony, were attacked by a Muslim mob from numerous nearby Muslim villages. Piecing together the story of what happened is difficult but, from what I gathered following the destruction, there was a central figure involved in the unfolding of the riot, a Christian male who combined illegal liquor sales with gambling (which is illegal for everyone).

The ongoing sustained double offense was made possible because this man paid bribes to some local Muslim police officers. On one occasion, when the police officers demanded what was probably a higher payment for their ongoing acquiescence, the Christian refused. The police then threw a bible onto the ground and stamped their feet on it. This behavior greatly offended the Christian and his cohorts, so they appealed to the local chief of police.

Sensing the symbolic nature of the offense, the police chief fired the police officers. Of course, this infuriated them. In their anger, they allegedly appealed to some local muftis, telling them that some Christians in Shantinagar had torn pages out of the Qur'an, crumpled them, and thrown them on the ground. They fabricated physical evidence to prove this, and showed it to the muftis. Outraged, the muftis announced from loudspeakers on the minarets of the mosques throughout the surrounding villages that the Muslim faithful should "go teach those Christians a lesson."

A mob formed and went en masse to Shantinagar and Tiba Colony, violently attacking them both. Fortunately, only one person died in this violence—an attacker suffered a heart attack. Houses were burned, and tractors, cars, trucks, and irrigation pumps were destroyed. Children were beaten. One Catholic high school and adjoining hostel, one high school and one grade school of the Salvation Army, and a total of thirteen churches were ransacked and torched. This is an example of what faith communities can do when they believe there has been a moral violation and they are urged to take action.

Closely related to this sense of being offended because of a moral violation is a view of "holy entitlement" when sacred space is involved, especially related to a perceived God-given responsibility or privilege to possess or control land. Scholars call these "sons of the soil" conflicts (see, for instance, Korf 2005; Toft 2005).

To be a sons of the soil conflict, one group involved believes that it has sacred entitlement to a specific place and a duty or religious mandate to hold on to the land. For example, the Middle East conflict is complicated because of a felt sense of entitlement by some of the Jewish people to the "promised land" (not to be confused with a perspective of many others who see Israel as a place of refuge after centuries of discrimination and slaughter, culminating in the Holocaust). Similarly, the civil war in Sri Lanka from 1983 to 2009 was driven in part by the belief that the island of Ceylon (its former name) must be kept predominantly Buddhist, as part of a religious mandate to preserve the "Golden Kingdom of Buddhism." As Rapoport explains:

So in either case, when there has been something sacred which has been dishonored, or when an entitlement is being encroached upon, a sense of sacred duty becomes operative. That is one reason why violence between groups of different religious identities can be so intractable. Militants of a religious group make public accusations or spread rumors that the group's faith has been defiled. They often make references to religious symbols, such as a disturbance of a holy ritual or the degradation of a hallowed site. (1993, 452)

An approach to use in overcoming this intractability is to engage in dialogue that involves what theologians call *hermeneutic variability*. This entails an exploration of alternative interpretations of sacred understandings. Or, as more comprehensively explained by Gopin, it is the broadest way in which stories, institutions, rituals, texts, precepts, and values are being reinterpreted all the time, seen in a new light by virtue of the interactions between the individual or the community and the environment and life history in which they find themselves. I assume that even those who believe religious beliefs and practices to be static or see themselves as changing only in order to revert to some idealized past structure are, in fact, engaged in subtle but very real processes of development. That . . . suggests that much more is changing than is generally realized, not that there is nothing consistent in a tradition, an argument that would be demonstrably contradicted by empirical evidence. (2000, 59)

It is perhaps natural for those of one faith to point a finger at those of another faith when aggressive statements or acts have occurred between them. But when one religious group criticizes another, tension often becomes greater.

A different way of communicating is offered by the Carnegie Commission on Preventing Deadly Conflict, which states that religious leaders should "take more assertive measures to censure coreligionists who promote violence or give religious justification for violence" (1997, 118). Essentially, the Commission advocates that confronting aggression will be most effective if done by those of the same faith, not by those who are recipients of the perceived or actual aggression. And this makes sense in that moderates of the same religion of aggressive extremists are intimately familiar with the sacred texts, doctrines, perceptions, and culture of their in-group.

However, moderates of the recipients of aggression should know the moderates of the aggressing group so that a concern about aggression is conveyed. This is required in part because aggression, like beauty, is "in the eye of the beholder."

The "architecture" of these streams of intrafaith and interfaith communication and interaction is depicted in figure 1.2. Note that religious leaders are only one of three kinds of leaders who can impact in-group and out-group communication and interaction. Civic and political leaders have important roles as well. Their critical attributes, generally speaking, are being respected and having credibility within their respective groups.

The engagement of moderates in counteracting a view that violence is justified—which is propagated by extremists of their own faith or ethnicity, or both—is often an overlooked focus of violence prevention. Reflecting on the March 2004 riots in Kosovo between predominantly Muslim ethnic Albanians and Orthodox Serbs, Chigas concludes:

Intra-ethnic social networks (or "bonding social capital") were more important than inter-ethnic engagement in preventing violence. Communities that avoided violence in March 2004 experienced no influx of "newcomers" and were generally able

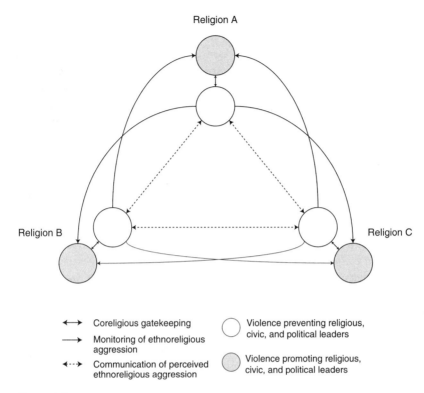

Figure 1.2
Interfaith and intrafaith communication

to bridge intra-community political divides. As a result, intra-ethnic social networks remained intact and strong and were a significant resource for dissemination of information and mobilization of collective action. Where communities have access to relatively reliable information about the other's intentions, and the situation, leaders anticipated the arrival of violence in their communities and took action to interrupt the cycle of action-reaction.[2] (2006, ix)

Why is religion such a potent force in promoting violence, on the one hand, as compared to fostering respect and reconciliation, on the other? Rapoport offers this cogent observation:

When a religious justification is offered for a cause, which might otherwise be justified in political or economic terms, the struggle is intensified and complicated enormously. There are many reasons why this happens, perhaps the most important being that religious conflict involves fundamental values and self-determination; and struggles involving questions of identity, notoriously, are the most difficult to compromise because they release our greatest passions. (1993, 446)

Emotional escalation because of a sense of moral violation often uses religion while not being based on, necessarily, religiosity. For instance, observers of the war in the Balkans witnessed Croatian fighters carrying rosaries
(consisting of a cross and beads on a string or chain used for saying a series
of prayers, a practice of Catholics), but few of the young soldiers carrying
them actually knew how to pray with them. And this ignorance is merely
a symptom of a larger problem of what Appleby calls *religious illiteracy*
which feeds into violence. Here is how he describes it:

The low level or virtual absence of second-order moral reflection and basic theological knowledge among religious actors—is a structural condition that increases the
likelihood of collective violence in crisis situations. . . . A supremely self-interested
and skilled politician or preacher confronting—or having assembled—a mass of
people outraged by their "victimization" at the hands of ethnic or religious others
may easily exploit deep emotional currents and volatile prejudices in an audience
drawn from a religious but religiously illiterate population. . . . The selective retrieval
and politically motivated interpretation of one dramatic episode from a vast,
complex, and ambiguous history . . . serves to construct and demonize "the other,"
to solidify and channel extremist passions, and to extend a sacred canopy over the
whole dubious process. (2001, 69)

One often hears that conflict over religion is really not religious conflict.
Rather, it is conflict between groups that are suffering from xenophobia
and fear that one group will become violent toward another group. Yes,
religion is a factor but the aggression is not an inherent imperative of a
given religion. Its moral strictures and social teachings are simply presented so that aggression seems required. The real struggle, in fact, is not
an interfaith one; rather it is more aptly viewed as primarily an intrafaith
challenge. "Religious conflict" is not mainly, for instance, between Muslims,
Hindus, and Christians. It is substantially conflict between moderates and
extremists of the involved faith communities.

   We must engage, however, rather than shun orthodox religious leaders,
some of whom are knowingly providing rhetorical justification for extremist militancy, and others who are doing so inadvertently. When moderates
confront coreligionist extremists, skillful use of cognitive dissonance might
be a helpful tactic within a context of respect (not as manipulation). The
theory of cognitive dissonance was developed initially by Leon Festinger
(1957). Festinger and subsequent psychologists postulate that when a
person holds thoughts that are in a dissonant state, the cognitive discomfort results in unconscious or conscious processes seeking to obtain consonance. It is comparable to hearing dissonant sounds in music which
create a sense of unease, as one typically encounters when watching a

suspenseful part of a play or movie, waiting with anticipation for the crescendo of an act that will reestablish a consonant state.

In the case of intrafaith communication between moderates and extremists, creating cognitive dissonance through dialogue can be a powerful force (Steele 1998). This can be done by pointing out a major thrust of the religion as being inconsistent with an extremist interpretation. Holland, Meertens, and van Vugt identify a subset of cognitive dissonance. They find evidence of what they call *moral dissonance*—what a person feels when doing harm to others. As one might expect, they also find that those with low self-esteem tend to engage in greater self-justification than those with high self-esteem, which suggests that engaging in dialogue with extremists who have higher self-esteem is likely to be more efficacious than engaging with those who have lower self-esteem (2002).

When private engagement is unsuccessful, public dialogue or debate might be warranted in that it can potentially convince reluctant followers (or simply curious bystanders) that violence is not a moral imperative. I have written elsewhere that hermeneutic dialogue sometimes needs to be public—even if it involves heated exchange—when moderate leaders feel compelled to confront militant, intransigent coreligionists; I refer to such dialogue as "theological dueling" (Bock 2001a; Bock and McCauley 2003). I presented that concept while teaching a class along with Fr. Cedric Prakash (former director of St. Xavier's), on religious leadership and violence prevention. He argued that "dueling" is an inappropriate and inaccurate label. I did not understand his point, other than to recognize that "dueling" is a didactic term. In my experience and research, I have found that public discourse about the meaning of sacred texts among those of the same faith tradition can be vicious. It usually is not dialogue in the sense of polite socializing. It is far more challenging and, for many, unnatural. And to be effective, one needs to intervene early. Doing so, however, can appear alarmist. It is unpleasant to confront a coreligionist who seems to be starting down the road of *xenophobic religious nationalism* (a condition existing in situations in which leaders of a group use religion to claim superiority of their group relative to other groups, marked by close-mindedness to other perspectives) while knowingly or unknowingly giving power-hungry, sociopathic troublemakers or spoilers ammunition for religiously "justified" violence. To me, this is comparable, but even more difficult, to confronting one's brother-in-law at a holiday meal after he says an inappropriate joke. It simply is not pleasant.

But as I later pondered Prakash's caution, I pulled my copy of Paulo Freire's *Pedagogy of the Oppressed* from my bookshelf to review his description of respectful engagement. I found, among others, this insightful passage: "Dialogue cannot be reduced to the act of one person's 'depositing' ideas in another, nor can it become a simple exchange of ideas to be "consumed" by the discussants. . . . it must not serve as a crafty instrument for the domination of one [person] by another. The domination implicit in dialogue is that of the world by the dialoguers; it is conquest of the world for the liberation of [people]" (1997, 77).

Respectful dialogue is not manipulative. But one should not ignore psychological dynamics at play, including cognitive dissonance. When people hold two differing views at once, they feel a dissonant cognitive sense. Hermeneutic dialogue can navigate beyond cognitive inconsistency to respectfully probe for truth by juxtaposing extremist views with major thrusts of a particular religious tradition. David Steele observed this happening during the horrific violence in the Balkans following the downfall of Tito and the subsequent disintegration of the Yugoslavian nation-state that was marked by ethnic cleansing in Kosovo and elsewhere. Steele observes, "One could simply dismiss as manipulative posturing an attempt by a Serbian Orthodox priest to both espouse nationalist ideology and call for humane treatment of Muslims and Croats. Or one could understand the influence of the apostolic tradition of the Orthodox church on him and, therefore, recognize the priest's struggle with unreconciled parts of himself" (1999, 23). A respectful coreligionist can sense this dissonance and engage in dialogue to explore its meaning. The drive for consonance creates an opportunity for learning and collective discovery.

A Theory of Change from Public Health

Many in global health see collective violence as being a result of a societal disease. Gary Slutkin, for one, regards collective violence in this way. A physician at University of Illinois-Chicago, Slutkin worked for a decade in Somalia, Uganda, Thailand, Kenya, Tanzania, Malawi, Congo, Zaire, and many other countries to eradicate infectious diseases (mainly HIV/AIDS and tuberculosis). When he returned to the United States in 1995, it occurred to him that gang violence in Chicago could be addressed as an epidemic, similar to how he addressed disease in Africa and Asia. He and his colleagues developed a program called CeaseFire, which is explained in detail later in this book. Here, we cover its underlying theory.

CeaseFire was evaluated by a team led by Wesley Skogan at Northwestern University (Skogan et al. 2008). The evaluation team was struck by

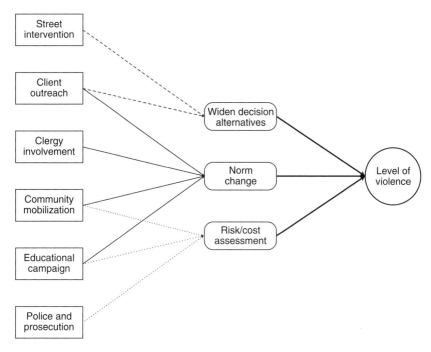

Figure 1.3
CeaseFire's theory of change. Permission to use this figure was provided by Wesley G. Skogan.

a well-articulated "theory of change" integral to CeaseFire's approach. Figure 1.3 is a diagram of that theory.

It is noteworthy, as explained previously, that Horowitz identifies the reduction of risk as one part of the consensus-building process in his analytical framework. Similarly, Lederach emphasizes the importance of having options, or alternatives. The CeaseFire "theory of change," however, adds another important component—the norms influencing a social group, that is, the "beliefs, attitudes and values that make up a culture of a community" (Skogan et al. 2008, 1–5).

CeaseFire's theory identifies what needs to happen to bring change about (they call these catalysts "levers"), including direct personal contact ("client outreach"); the engagement of those with moral authority ("clergy involvement"); the mobilization of the community; and conducting an educational campaign. The criminal justice system creates a sense that those who become violent will potentially spend time in jail, thereby increasing sense of risk.

Some Disclaimers

In terms of testing Horowitz's framework, this book does not focus on whether all of the components of consensus building are present before violence. While it would be useful to test empirically this short-circuiting hypothesis (that counteracting one of the processes that occur during a lull will prevent violence, rather than a combination of the five processes, as depicted in table 1.1), I do not have adequate data to do so.

The data I was able to acquire for the quantitative analyses reported in this book are from CEWARN in the Horn of Africa and from FCE in Sri Lanka. Results of how long the lull typically lasts are presented, which provide an indication of who has a reasonable chance of intervening to prevent violence. Also, this book covers the extent to which efforts to mitigate violence are effective after there have been precipitating events.

This book does not analyze violence prevention relative to Horowitz's framework as it might relate to conflict writ large (that is, *macroconflict*). For instance, military units often are given orders to attack on very short notice. There is virtually no lull or consensus building. Rather, there is imbedded automaticity related to military discipline inculcated through extensive drilling. The violence analyzed in this book is the type that civilians, not belligerents of standing armies or rebel groups, are involved in waging.

Keep in mind that Horowitz developed his analytical framework for ethnic riots in particular, not violence at a local level in general. It is an open question whether localized violence in general will have a precipitating event followed by consensus building which results in violence. It is reasonable to assume, however, that this type of sequence is common: local conflict involves a consensus-building process, certain events trigger those processes, there is usually a delay during which consensus is built, and then there is violence, depending upon the type of consensus built.

This book addresses conflict at a local level in general. I have not sought to analyze conflict patterns more specifically within the various subgroups of local conflict, though I concur with scholars who contend that we have considerably more work to do in developing a categorization scheme for local violence. Research into the length of the lull for different kinds of local conflict will add greater precision to understanding violence at a local level. For instance, ethnic conflict might show a slightly longer lull than, say, ethnoreligious conflict when religious leaders engage in demonizing those of an out-group.[3] And gang violence in Latin America or U.S. inner cities, for instance, is likely to have different dynamics as well.

I differ with some scholars in my epistemological perspective. To me, assessing the usefulness of analytical frameworks and theories and making adjustments to them are part of reflective practice. As Schön points out, this involves assessing "messy, problematic situations . . . implicit in the artistic, intuitive processes which some practitioners do bring to situations of uncertainty, instability, uniqueness, and value conflict (1983, as quoted in Cross 2001, 53–54).[4] While I appreciate the contributions made by the disciplines of psychology, sociology, social-psychology, anthropology, political science, political psychology, political geography, and economics, I consider them akin to what "pure sciences" like physics and chemistry are to engineering and architecture, the latter of which are applied sciences that require their practitioners be "reflective." Dealing effectively with violent conflict demands an understanding of how violence unfolds so that we can design and implement approaches to counteract it.

While I am appreciative of the work of those who strive diligently to develop taxonomies of conflict at a local level, the focus of this book is more about how violence unfolds, how we can detect that it is unfolding, what can be done to stop it, how much time there is to stop it, and who can reasonably be expected to take the actions needed to stop it within that time frame. I anticipate that additional work in "pure social sciences" will yield more precise categories. It will refine our understanding about the conditions and events that lead to violence within those categories, and elucidate patterns of how that violence unfolds. But I do not feel researchers and practitioners focused on applied theory and practice need to wait until that research is further along to design approaches that are useful in preventing the violence causing massive human suffering all over the world. I suspect those involved in "pure social science" feel similarly.

Summary

In this book, we seek to sharpen our understanding of how deadly violence unfolds by using insights from Horowitz's analytical framework, perspectives about the greed-grievance nexus, iniquity theory, the sons of the soil problem, cognitive dissonance theory, and CeaseFire's theory of change. In embracing these, we formulate an "applied theory of violence prevention" for reflective practice that has these dimensions: underlying conditions, time and space considerations, pathological social-psychological processes, and "levers" for impact. Of the *underlying conditions*, we include a history of enmity, a precipitating event, and norms. Of the *time and space considerations*, we focus on the length of time between a precipitating event and the onset of violence, the lull, and seek to address how long people have to intervene when warning signs are recognized after a precipitating event. We also consider the proximity of where interveners need to be relative to the soon-to-be attacking group in order have a chance to counteract successfully consensus building to become violent. Of the *pathological social-psychological processes,* we include emotional engagement, justification for killing, exaggeration of threat, reduction of risk, the lack of options, exploitation of grievances by militant leaders, and a desire to maintain honor and to fulfill a sacred duty. Among the *"levers" for impact* we include outreach (by those with whom the group can relate), community engagement, community education, moral leadership, dialogue involving hermeneutic variability (often involving a struggle to overcome cognitive dissonance), involvement of police officers or peacekeepers, or both, and prosecution of perpetrators (especially of spoilers and troublemakers).

2　Reporting and Warning about Deadly Possibilities

Recognition is the first essential in a chain that leads to the prevention of predictable surprises. Once recognized, however, leaders must make preventive action a priority in the midst of a noisy environment rife with competing priorities. The potential for emerging threats to be recognized but not prioritized is considerable.

—Max H. Bazerman and Michael Watkins, *Predictable Surprises: The Disasters You Should Have Seen Coming and How to Prevent Them*

The capacity to generate patterns, when harnessed to the potential of emerging ICT . . . presents the humanitarian and early warning communities with new assessment possibilities in real and actionable time-frames.

—Jennifer Leaning, "The Use of Patterns in Crisis Mapping to Combat Mass Atrocity Crimes"

Early warning and early response for violence prevention involves three main activities. First, those monitoring a conflict must collect information. Second, the information must be analyzed over time. And, third, warnings must be conveyed to those who can take action. Of course, if a warning does not result in a response, then these three steps are of no value.

In this chapter, we cover how information about conflict and cooperation is collected, how it differs from more static indicators, and how it is weighted. We also assess the accuracy of developing data using automation as compared to creating it manually.

Early Warning Compared to Risk Assessment

When my family and I agreed to move to Pakistan with Catholic Relief Services (CRS), we had to make a number of changes in our lives. One challenge we faced was in determining how to manage our finances. We decided to open a discount brokerage account. The reason I mention this is that I inadvertently learned about quantitative analytical approaches in the process. These approaches are very similar to what I have since learned about risk assessment and conflict early warning.

Every month while working overseas we received a newsletter from our brokerage company. It profiled stocks and presented a recommended portfolio. My wife Sue and I had tried to keep our Individual Retirement Account savings in socially responsible mutual funds, but had been disappointed, generally, with their performance. However, after reading the Standard & Poor's newsletter for a few months it seemed like it would be a simple innovation to buy stock recommended by the newsletter and to sell it when a subsequent newsletter showed it was no longer in the portfolio. The only research I needed to do was to see if the company's products and services were consistent with our ethical and moral considerations.

This proved to be a simple and workable approach for us. I had to spend minimal time every month selling those stocks deleted from the portfolio, and screening and buying the new ones. But as human nature would have it, I became curious as to the reasoning behind the recommendations to buy or sell. It was then that I started learning about *fundamental* and *technical indicators*.

It was a bit overwhelming at first to learn the investment jargon. But as I read the reports of individual companies online, it became clear that fundamental indicators are those that are relatively static. Fundamental analysis provides an assessment of the overall "health" of a company, based on items such as the amount of debt a company is carrying, the price of the stock relative to the earnings of the company, and the value of assets.

Technical indicators, in contrast, relate to recent trends, such as in the value of the stock (as compared to its past value) and the turnover in buying and selling it. A very basic pattern that many stock investors use is a "cup and a handle." When a stock's price dips but then goes up (called "support") forming a cup, some investors look for a temporary period of the price going down again (called "resistance"), waiting for it to break out of that brief downward price period by going up again (called "building a base"), which is the "handle." This is seen as verification that the stock is on an upward trend rather than going up and down like a ping-pong ball.[1] The "cup and a handle" is but one example of patterns that are recognized by stock traders to be an indication of a buy opportunity.

Pattern recognition sounds complicated, but think of it this way. Imagine you are in Venice, sitting at a sidewalk café waiting for a good friend (not a bad place for a rendezvous!). You see various people walking toward you on the road but you spot your friend before you can discern his or her facial features. You recognize his or her walk, for example, or distinctive posture. That is pattern recognition. In finance and other areas of analysis it can involve movements, graphs, or mathematics, or a combination. It is a way of identifying something based on what we can discern now for the purpose of making a decision. For example, when you are sitting at the café and recognize your friend, after you'd thought it was time to leave, you may decide to stay longer, to wave, and to call out a greeting.

In assessing the likelihood of violence, researchers have made a distinction between risk assessment (which is like stock traders' fundamental analysis) and early warning (which is like stock traders' technical analysis). One of the most widely known risk assessment projects was undertaken by the State Failures Task Force (Murshed 2008, 1). The use of risk assessment to predict violence has made major strides thanks to pattern recognition software. O'Brien uses a specialized algorithm called Fuzzy Analysis of Statistical Evidence (FASE) that successfully predicts country instability with 80 percent accuracy five years out (2002). He refers to fundamental indicators as "oily rags." Using that analogy, a precipitating event is a "spark." To O'Brien, "the oilier the rags, the more likely a single spark (i.e. riot, natural disaster, or assassination) might produce an explosive situation" (796).

Events are relatively unpredictable. They can cause dramatic shifts in the behavior of groups suffering from substantial intergroup tension. The structural causes of conflict, such as discrimination over decades or centuries, are a foundation upon which "brush fires" of violence can ignite. But it would be a mistake to use an approach for knowable processes that involve the "big picture" (the "forest") for generating an early warning that relates more closely to relatively recent events (the "trees"). Yiu and Mabey warn against using a "tree" approach at a "forest" level. This is because a "deep-but-narrow focus will often overlook the generic relationships driving instability risk." In contrast, using a "forest" approach at the "tree" level can result in "misleadingly strong and unpredictably biased assessments" (2005, 10). Claiming that a country is about to deteriorate into massive violence based on bloodshed in the capital can be deceptive. Risk, they argue, can be assessed productively with "knowable processes" in which "cause and effect are separated in space and time." On the one hand, the application of research and analysis regarding risk can yield improved understanding which can inform policy decision making (Yiu and Mabey 2005, 9; Kurtz and Snowden 2003). Early warning, on the other hand, is a better approach to use with "complex processes." These are characterized by incoherent cause-and-effect relationships that are better understood after events unfold. Gaining clarity about them "requires the use of 'probes' or actions to stimulate system responses, detailed monitoring of emerging patterns and flexible responses" (Yiu and Mabey 2005, 9). The data used for early warning are *dynamic*. And the "probes" are field officers of NGOs, members of peace committees, or local observers reporting events via radios, cell phones, social media, or online forms.

Using events data for early warning of conflict is possible largely due to a greater computation capacity. As Schrodt and Gerner point out, also drawing a parallel with approaches used by stock traders:

Until relatively recently, technical stock market analysis generally had a bad reputation, due to its use of statistically dubious patterns based on small samples, wishful thinking, and gurus whose fortunes derived more from the sale of books than from trading stock. With the increase in computing power in the 1980s, the situation changed, and "programmed trading systems" can now process sufficiently large amounts of information to generate profits working solely with information endogenous to the market itself. The greater information processing capacity today, in contrast to that available in the 1970s, may have a similar effect on the analysis of political behavior using events data. (2000, 805)

Clearly, potential violence must sometimes be analyzed by investigating both knowable and complex processes. For instance, if a risk assessment approach points out that a particular country is likely to become violent, it may be helpful to put into place an early warning system as a way of staying diligently informed about potential flare-ups.

While structural analyses use indicators relating to governance, inequities, poverty, and economic conditions, most early warning systems have focused on events of cooperation and conflict. More recently, however, structural data are being analyzed along with events data, thereby improving predictive capacity (Meier, Bond, and Bond 2007; O'Brien 2010). swisspeace, a think tank based in Bern, Switzerland, for instance, linked its "Early Analysis of Tensions and Fact-finding" system (known by its German acronym as FAST) with a risk assessment model that uses structural data developed in the Facts on International Relations and Security Trends (FIRST) project at the Stockholm International Peace Research Institute (SIPRI).[2] Therefore, the distinction between risk assessment and an early warning systems is being blurred, again not unlike some stock-trading advisory services that combine fundamental with technical analyses.[3]

Generating Events Data

Events data are nothing other than coded information about what happened—about discrete incidences of conflict or cooperation. For example, assume that we are building an events data base in Burundi, where there has been considerable tension between Hutus and Tutsis, especially following the genocide in Rwanda. Assume you learn that someone from one ethnic group is harassed by people of a different ethnic group while going through a checkpoint. You would create events data by coding the identity of the victim (let's say "T" for Tutsi), the identity of the perpetrator (let's say "H" for Hutu), the type of conflict (let's say "HARR" for harassment), the date, and the location (using, for instance, GIS coordinates).

Now, assume later that day a Tutsi politician is assassinated by a Hutu in the same location. The data would be identical except for the code for the type of conflict (which would be something like "ASSA" for assassination).

Violence prevention systems that use events data generally follow a scheme for categorizing cooperation and conflict on various scales. One example is the Integrated Data for Events Analysis (IDEA) protocol. Formulated by Bond, Bond, Oh, Jenkins, and Taylor (2003), it consists of over two hundred event types of both conflict and cooperation. IDEA's main categories of cooperation are agree, consult, endorse, grant, promise, reward, yield, and other. Its main conflict categories are accuse, criticize or denounce, complain, demand, demonstrate, deny, expel, force use, reject, sanction, seize, and threaten. Main categories sometimes have subcategories. For instance, the main category of seize has these subcategories: abduction; arrest and detention; covert monitoring; seize possession; and seize, not specified.[4]

While IDEA and other similar schemes provide a helpful beginning list of events, it is not uncommon to adapt them by adding site-specific categories. For instance, in a workshop in Colombo, Sri Lanka, with both field and headquarters staff members of the Foundation for Co-Existence (FCE), Dominic Senn of swisspeace and I were told that the following categories needed to be added: (1) incitement by religious leaders; (2) politicians urging people to launch a *hartel*, a shut-down of businesses that sometimes become violent, as a form of protest; and (3) accusations of favoritism during post-tsunami relief operations. Interestingly, they also asked that they be able to link one event to another in the database since doing so would help them identify patterns and organizational involvement. And, finally, we agreed that the system should require a preloading of the contact information of local-, middle-, and top-level moderates to facilitate contacting them quickly in a crisis.

Others with early warning and early response experience stress the importance of incorporating events related to gender. Female leaders who have a stake in the outcome of the conflict are often influential in violence prevention, so their actions should be incorporated into whatever categorization scheme is used (Nyheim 2008, 16; Schmeidl and Piza-Lopez 2002).

Still others who work closely with people locally stress the importance of asking which events are most important. The firm Virtual Research Associates (VRA) is supporting the development of three early warning and early response systems in Africa, none of which use preconceived categories. For each system, VRA has convened workshops to develop lists, and additional workshops for "calibration" of each event type, another way of saying that events of greater importance are given a higher weight.[5]

Automation and Accuracy

A disadvantage of coding events by hand is that it is labor intensive. Another way to generate events data is to use a computer program. This is generally called an *automated* approach. Automated events data generation uses artificial intelligence techniques to distill data electronically from news stories, web posts, radio broadcasts, and public pronouncements.

How does an automated approach work? We must first distinguish between *information retrieval* and *information extraction.* Information retrieval is similar to a bibliographic search using key terms typed into a library collection database. Once key words have been entered, the computer looks for articles and books that contain those words either in the title or in the text itself, or both, depending on instructions given or limits of the software. Or, more simply, when opening a document using a word processing program, one types in key words, waiting for the computer to show the files that have those key words in the title or, in the case of an "advanced" search, in the body of the document.

In contrast, information extraction dissects text into specific parts, discards what is not useful, and keeps what is useful in a form that can be analyzed. This is often done in three stages: *tokenization and lexical processing, syntactic processing,* and *domain analysis* (King and Lowe 2003, 638–641; Hobbs et al. 1992). Tokenization and lexical processing involve dividing words into their grammatical categories (such as nouns, verbs, adjectives, and adverbs), with an ability to handle fuzzy logic. Software that can handle *fuzzy logic* is capable of "learning" associations over time. For instance, software capable of handling fuzzy logic can recognize that the word "Bob" is often associated with "Robert." It can also distinguish different meanings of "bank": a place to deposit money, for example, or where the land meets a river. The software identifies an association with financial transactions with the former and with water and shoreline with the latter type of bank (Zadeh 1965; Hobbs et al. 1992, 7; King and Lowe 2003, 638).

Syntactic analysis interprets data that have been "tagged" during the tokenization and lexical processing phase. During this phase, the computer looks for additional information in the document as a way of gaining greater clarity on the meaning of specific words. For instance, fishing for information about secret banking accounts during an interrogation session is very different from fishing for trout on the bank of a river. The computer must be able to distinguish these two uses of the word "fishing." Linking the first use to "interrogation," the computer would associate "fishing" with a specific type of conflict event as compared to a recreational activity.

Another example of syntactic analysis is when a computer identifies the difference between a fight of a married couple and a fight between armies as a part of a border dispute. The data gathered on each event include *who did what to whom, where, when, and how.* In the case of a marital dispute, the "who" of "married couple" or "husband" or "wife" would disallow the event to be added to a database on collective conflict since the software is looking for, for instance, "ethnic group" or "rebel group" or "insurgents" or "country." Syntactic analysis is sometimes called *parsing.*

Finally, in domain analysis the computer seeks to eliminate ambiguity in situations where the meaning of a word has been inconsistent. For instance, the computer must learn to understand that the "Liberation Tigers of Tamil Ealan" in Sri Lanka is the same as "LTTE" and the same as "the Tigers."

One way that the speed of information extraction is enhanced is by identifying low-frequency "trigger words" that are sometimes referred to as "rare events." Depending on the intensity of a conflict, the word "hostage" or "missile" would be identified and the sentence structures related to those "trigger words" would be the main, if not the only, focus for information extraction.

You might be wondering about the accuracy of events data produced through automation. Schrodt and Gerner (1994) tested the accuracy of machine-coded events data for five Middle Eastern countries, the Palestinian Territories, and the United States from 1982 to 1992. The machine coding program they used was the Kansas Events Data System (KEDS), developed at the University of Kansas. The machine coding program uses Reuters newswire leads as the source of information for generating events data. For comparison, data from the *New York Times* and the *Los Angeles Times* were compiled into World Events Interaction Survey (WEIS) categories by people who were knowledgeable in coding events data. The WEIS categories were developed by McClelland (1972) for international relations research (that is, at the macrolevel). Data on international cooperation and conflict were developed by human coders and made available to researchers of international interactions from 1966 to 1992 (Schrodt and Gerner 1994, 826).

Schrodt and Gerner found an 80–85 percent correlation between machine-coded and human-coded events (1994, 830). They concluded that the machine-coded data "identify the major trends of international conflict and cooperation in the region . . . specific patterns over time are generally consistent with the narrative record . . . major events . . . show up clearly in the data" (851). They did, however, find that data for some of the interactions between countries—namely, those which involved Lebanon and Jordan—were less than anticipated.

In another, more recent, study, King and Lowe (2003) compared human-coded with machine-coded data generated using Virtual Research Associates' VRA Reader. Specifically, they trained three undergraduate students and had them code events. At the same time, they used the VRA Reader to code events from the same news articles in the Reuters Business Briefing. The Reader begins by reading the first sentence, or lead. It codes events into 157 IDEA categories.

King and Lowe then evaluated the performance of undergraduate coders and the VRA Reader relative to an accurate coding of the same events by experts. They first calculated the accuracy of events coding into subcategories, then the accuracy of those data when they were combined into the major categories and, finally, the number of coded events as compared to actual events. For subcategories, King and Lowe concluded that "the Reader places an event in the correct one of the 157 [event] categories 26% of the time . . . [whereas] the three undergraduates were correct, computed in the same way, 32%, 23%, and 26% of the time, respectively. For this particular method of evaluation, the Reader thus falls squarely within the range of our human coders" (2003, 631).

This degree of inaccuracy is disconcerting—both of the machine and the undergraduates. When King and Lowe combined the data into the main categories, however, they found that the Reader and the undergraduates were accurate 55 percent, 55 percent, 39 percent, and 48 percent of the time, respectively, again concluding that there was little difference between the performance of the machine as compared to the undergraduates (2003, 631–632). So, in the aggregate categories, they were accurate roughly half the time.

In a third measure of performance, King and Lowe tested the extent to which the machine and the undergraduates generated events when they happened or, conversely, generated events when they did not happen. That is, they measured instances of events either of conflict or cooperation, not type of events. In this instance, they found a remarkable degree of accuracy—93 percent, 94 percent, 80 percent, and 90 percent for the machine and three undergraduates, respectively (2003, 632).

What about the trend lines of overall conflict and cooperation? This is the *actionable information* that an early warning system needs to provide (that is, data, often depicted graphically, with which to determine if violence is likely so as to decide whether to intervene). In this case, King and Lowe concluded that points plotted on the trend line by both the Reader and the undergraduates "cluster fairly closely" to the trend line produced by the expert coder, with a slight bias (of both) in underestimating the degree of conflict (2003, 633–634).

In summary, even though the coders assigned the incorrect codes to roughly three-fourths of the events, they got the major categories right about half the time and generated events when they happened with considerable accuracy. Taken together over time, these results suggest that the overall trend toward conflict or cooperation was "good enough" for the purpose of early warning.

There are at least two ways of looking at the view that "approximate accuracy" is "good enough." One is that with so many events being generated, precision is less important than the pattern. A different view is to compare it, as before, with the analysis of stocks. The pattern witnessed in a time series of a stock's price is often overwhelmingly influenced by momentum, that is, the volume of stock being traded as compared to its "normal" volume of trade—a technical indicator. When it is clear that momentum is well above average, look out, because there is a stronger than normal likelihood that the price of the stock is likely to fall or rise dramatically (those who track these patterns and make short-term investments are sometimes called "momentum investors," trading either "long"— assuming the price will go up, or trading "short"—assuming the price will go down).

One way to enhance accuracy of warnings is to incorporate other information that has an impact on tension a population is encountering, some of which is readily available. For instance, Meier, Bond, and Bond (2007) found that combining environmental data, particularly on rainfall, with events data on conflict and cooperation increased the predictive capacity of CEWARN.

There is a sense in which the architecture of an events data system is helpful despite the accuracy of the data and the validity of the reports. Put another way, striving for accuracy and reporting with regularity can have positive outcomes despite unreliable data thereby produced.

When field staff members are involved in the creation of a categorization scheme on conflict and cooperation, and they are required to submit reports on a regular basis using that scheme, they develop a frame of reference that has greater precision in viewing conflict and cooperation across a broad spectrum. During interviews with FCE field officers, it was evident that being trained in how to use a categorization scheme, and then using it, had sharpened their analytical ability, though the data itself, or analyses derived from it, were not always mentioned per se. The FCE field officers indicated that their training in using the categorization scheme made them think differently. Somewhat surprisingly, however, Lawrence, Gaasbeek, and I found that there is an age bias in how FCE field officers viewed the value of data collection and use. Older field officers tended not to find the categorization scheme useful, while younger ones did. Therefore, the observation that using a categorization scheme helps users refine their analyses seems to be the case with younger workers especially (Bock, Lawrence, and Gaasbeek 2009).

The design of an early warning system also trumps data accuracy relative to levels of intervention. Peace committees operating on their own often lack access to mid- and top-level leaders. When a warning is generated at a local level (through inductive reasoning combined with mathematical pattern recognition, for instance) and then relayed to mid- and top-level leaders, a response is facilitated (though by no means certain). The "web of relationships" at all levels of society is critical, probably more so than events data accuracy or mathematical pattern recognition with those data.

The advantages of generating events data through automation are cost savings, reproducibility, and less influence of preconceptions and associated biases of coders (Schrodt and Gerner 1994, 827–829). But if an early warning and early response system involves field officers collecting information that is not in the media, then they are arguably indispensable, for three reasons. First, media syndicates typically do not report many of the subtleties of what is happening at a local level. This includes rumors, expressions of fear and concerns about threats, and other social psychological processes that are indicative of the kind of consensus building Horowitz (2001) identified.

Second, there is sometimes an advantage of local understanding that outside reporters do not have, which makes human data collection at a local level important. A good example that I heard, though it relates to international, macroconflict, was presented at the Oxford Futures Forum at Templeton College, Oxford, in October 2005. The presenter emphasized that sometimes people need to be inventive in identifying salient data. He gave this example: when there is speculation that the United States is planning military action, a helpful bit of information is a surge in pizza deliveries to the Pentagon—which suggests that people are spending long days and evenings working on battle plans.

A third reason why it is important to have staff members in the field is that they can intervene. Even if they themselves do not need to intervene, they can be critical in preventing violence by urging that local peace committee members become engaged.

Assigning Weights to Events Data Categories

A political assassination is much more likely to ignite widespread violence than, say, harassment of a worker at a checkpoint. How does one capture different magnitudes of severity? The answer: by assigning different weights to events categories. Think of it this way: how many instances of worker harassment (coded "HARR") would equal one political assassination ("ASSA"). Assume the answer is fifty instances of harassment would equal one political assassination. So, if the weight assigned to HARR is 1, the weight given to ASSA is 50. (This, again, is an example of when the exact weight is less important that the order of magnitude involved.)

Where do the weights come from? There are three ways of determining weights. One is to ask people (or groups of people, as with a focus group) what they view as most troubling, and how much more troubling one kind of event is compared to another. Or, in determining the weight for instances of cooperation, by asking how much more potent a cooperative event is as compared to another event, such as seeing Hutus and Tutsis playing together on the same sports team compared to Hutu and Tutsi political rivals eating dinner together publicly. This kind of consultation to determine weights can be done in any number of ways, including in person, over the telephone, and on the Internet.[6]

Another way to determine weights is statistically. But using that method requires enough events data to determine probabilities with findings that are statistically significant. Assume you work for swisspeace. You and your colleagues gather qualitative information and events data from local area networks (LANs) in many countries. These networks are made up of one or more area experts, usually nationals, who live in the country under analysis. They submit reports on cooperation and conflict regularly. Their events data can be submitted on-line directly into a database, or they can send their reports via e-mail, for instance, to a central location where the data are entered. Over time, the data can be analyzed to see what sorts of events are most influential in causing violence.[7]

A third way of assigning weights to events data relates to those approaches that use crowdsourcing. Ushahidi, a crisis-mapping organization developed initially in Nairobi, Kenya, provides an online platform of events on a digital map. Those who manage the platform add greater credence to events that have been reported more than once, on the one hand, and close to the incident, on the other. For Ushahidi, reports that are duplicated or that change slightly over time (as in a "cascade" of data) are deemed to be more credible. Ushahidi developers are experimenting with a clustering method to depict events that are corroborated over time that they believe will be made more accurate by "filtering out fabricated reports." They call this filter their "digital straw."[8] The digital straw is part of what Ushahidi refers to as its "swift river," their approach to determine the accuracy of events data rapidly. Swift river involves pulling together streams of data that are "filtered through both machine based algorithms [that is, the digital straw] and humans [through crowdsourcing] to better understand the veracity and level of importance of any piece of information."[9]

Issuing Reports and Warnings

Moving into the second and third major components of early warning and early response systems, events must be *analyzed* so that *reports can be generated* which, ideally, provide accurate "snapshots" of conflict and cooperation. Analyses must bring order out of chaos by integrating otherwise unintelligible floods of information into a coherent picture. Reports must convey that picture with parsimony. They typically depict major trends in conflict and cooperation in specified locations. They do not usually provide warning because "sounding the alarm" is confined to a smaller audience.

With systems that do not collect events data, NGO staff or peace committee members, or both, go from one place to another to piece together disparate information and, through induction, develop an assessment of overall conflict and cooperation in their area. Sometimes these observations are entered into periodic newsletters that are sent to mid- and top-level moderate leaders and supporters.

Incipient violence often becomes most evident when rumors are circulating. In India, rumors often are spread systematically with the use of handbills (brochures), distributed widely, usually by youth or women, for hire for a rupee or two. A handbill might say, for instance, that Hindu youth from a nearby slum were seen throwing rocks into this slum of Muslims with the intent to disrupt celebrations of breaking a Ramadan fast.

An alert amounts to a telephone call to warn people that tensions are brewing. Sometimes peace committee members contact NGO staff members pleading for communication with moderate mid- and top-level authority figures who are disposed to intervene to prevent violence. In other instances, peace committee members themselves, trained beforehand in what to do in such situations, take it upon themselves to check the facts of a rumor and, if possible, discredit it.

With approaches that rely solely on human induction and also those that use events data, the basis of pattern recognition is essentially determining if something unusual is happening. Determining if a warning should be issued is a matter of tracking what is "normal" as a way of identifying the "abnormal." As Leaning notes, "Much of what we take to be knowledge or information is in the structure of 'compared to what' or 'what else is going on as well'. . . . Patterns—as opposed to lists of numbers or narrative accounts—allow us to see the full picture, as well as the details within it, literally at a glance" (2010, 205).

Trend analysis allows one to discern patterns over time, and these patterns are typically depicted in reports using various statistical methods and graphics. Descriptive statistics such as means, medians, range, and standard deviations can be used to provide an initial peek at how any particular variable behaves. Measures of *central tendency*, such as a mean (average), allow a single statistic to provide an impression of what a typical value would be among a set of data points. However, this measure possesses little utility unless we also know the *variability* seen in the original observations. The standard deviation is one statistic that gives us a sense of how much values typically vary about their mean.

When conducting trend analyses, some analysts might use a particular type of mean often called a *moving average*. One can, for example, add the values of the indicator of conflict for the past three weeks and divide this sum by three, and do this for each week for all week over a particular period of interest. One can then compare this *three-week moving average* to a similarly calculated *five-week moving average* (which is made up of the values from five days). A determination is then made as to how great a difference between the two moving averages constitutes a warning.[10] One approach would be to calculate the standard deviation for the five-week moving average at all time points, and then issue a warning any time the three-week moving average is one standard deviation greater than the five-week moving average—deemed to be an abnormal upswing in the aggregate conflict indicator (as seen in figure 2.1).

When the amount of information is sufficient, whether with relatively static indicators or with comparatively more dynamic events data, or both, techniques of greater sophistication for determining when to issue a warning can be used. For instance, Schrodt (2000) and Bond et al. (2004) have used hidden Markov models. Markov models are named after Andrey Markov, who developed a way of calculating probabilities of moving from one state to another. As Bloomberg and Hess (2002) explain, Markov models "can be thought of as dynamic contingency tables, in which the objective is to account for the observed transitions from one state at time period $t - 1$ to either remain in that state at time period t or to another at time period t."

Hidden Markov models are also used to make probabilistic predictions of change. One state, to give an example, might be "low intensity conflict" while another is "massive armed conflict." They are hidden to the extent that they are not directly observable. For instance, one cannot tell whether a conflict is massive or not when seeing isolated fighting. A model "learns over time" to identify when a movement will occur from one phase (or

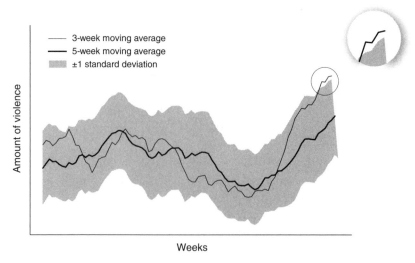

Figure 2.1
Moving average convergence divergence (MACD)

state) to another. By learning over time, I mean that as more events data are gathered, the pattern of when violence is likely to happen is refined as the data indicate, for instance, that a disruption of a sacred procession by those of a different religion is more potent in causing acute emotional reactions than the disruption of a political rally by those of another political party.

It is important to keep in mind, however, that hidden Markov models require substantial amounts of data. One expert, Joe Bond, estimates that they require approximately thirty-five events a day and can sometimes be a "black box."[11] By black box, I mean a mathematical approach that analyzes considerable amounts of data, yielding useful results, while at the same time the means by which those results are derived is inchoate. In the Eastern Province of Sri Lanka, the amount of data for each district during the country's prolonged separatist revolt from 1983 to 2009 was typically one or two incidents a day. A hidden Markov model would not work in that case. In contrast, using a hidden Markov model when receiving hundreds of reports daily (an amount received by large urban police departments, for instance) might be workable and useful.

More recent research has shown the value of combining structural data, events data, and analyses of statements of key leaders. Some scholars analyze proclivities of key actors by distilling their *operational codes*. An operational code is an axiomatic articulation of decision processes developed by identifying and analyzing statements made by key individuals (such as a commander of a paramilitary force) or small groups of people (such as leaders of a political party). O'Brien (2010) calls this an *agent-based early warning system*, meaning that it concentrates on one or more pivotal "agents" or leaders. Researchers decipher thought patterns that lead to specific types of behavior and decisions. This pattern recognition capacity, often using content analysis of statements made by decision makers, adds to the accuracy of warnings when using hidden Markov models.

Agent-based early warning systems are typically used at a macrolevel, focused on top political leaders. I am unaware of agent-based approaches being used at a local level, but I can see merit in using them with political or religious leaders, or both, in environments characterized, for instance, as having toxic levels of religious incitement.

Summary

Early warning differs from risk assessment. Violence can erupt quickly, while the conditions that create a sense of grievance that increase the risk of violence change gradually over time. Early warning relies on technical indicators that are dynamic. In contrast, risk assessment measures fundamental indicators that are relatively static.

Events data are created so that we can assess when violence is likely. Violence prevention systems that use events data generally follow a categorization scheme of cooperation and conflict. These data can be created manually or by using automation. The accuracy of events data is similar either way they are created—whether by people or by machine. But human coding is essential when events are not reported by media syndicates and are therefore unavailable for automated systems.

There is a sense in which the architecture of an events data system is helpful despite any inaccuracy of data. Put another way, striving for accuracy and reporting with regularity can have positive outcomes even when information is only approximate. When field staff members are involved in the creation of events data on conflict and cooperation, and they are required to submit reports on a regular basis, they develop a frame of reference that has greater precision in viewing conflict and cooperation across a broad spectrum of categories. They are also more inclined to seek out information rather than receive it passively.

Assigning weights to categories captures how some events are perceived as having greater significance than others. Regional variations demand tailored categorization schemes and associated weighting of events data.

Events must be analyzed so that reports can be generated that, ideally, provide accurate snapshots of conflict. Analyses must bring order out of chaos by integrating otherwise unintelligible information in a coherent picture. Reports must convey that picture with parsimony. They typically depict major trends in cooperation and conflict in specified locations. They usually do not provide warning because "sounding the alarm" is confined to a smaller audience.

With approaches in which events data are neither compiled nor analyzed, incipient violence most often becomes evident when rumors are circulating. An alert amounts to a telephone call from peace committee members to NGO staff members pleading for communication with moderate mid- and top-level authority figures who are likely to be inclined to intervene to prevent violence. In contrast, those systems that do compile events data typically identify trends. Some use mathematical approaches to identify acute changes in intergroup tension. From these calculations, warnings of violence are issued.

When the amount of data is sufficient, whether with relatively static indicators or with comparatively more dynamic events data, or both, more sophisticated techniques for determining when to issue a warning can be used. One approach is to use hidden Markov models that employ a form of artificial intelligence for pattern recognition. A drawback of this approach is that it requires large volumes of data.

Another method for early warning is to monitor the statements of key leaders to identify patterns of thought and threatening statements. This is called an agent-based approach, or otherwise referred to as analyzing the operational codes of key leaders.

II Violence Prevention on the Ground

Micro level responses to violent conflict . . . are an exciting development in the field [of early warning and early response] that should be encouraged further. These kinds of responses save lives.
—David Nyheim, "Can Violence, War and State Collapse Be Prevented? The Future of Operational Conflict Early Warning and Response Systems"

I'm a skeptic when it comes to how technology is used to support ICT4Peace [Information and Communication Technologies for Peace]. This isn't because of some dormant Luddite tendency, but because in the past ten years I've had high expectations about how technology could change the way in which the humanitarian community did business—and those changes have still not arrived.
—Paul Currion, "Conclusion"

In their insightful article, "The Electronic Oracle: Computer Models and Social Decisions," D. H. Meadows and J. M. Robinson seek to clarify the limitations of using computers (what they call an "electronic oracle," a source of knowledge or wisdom) to enhance decision making and improve public policy (Meadows and Robinson 2002). More recently, Thomas W. Malone expresses convincingly our challenge in understanding the limits of human–computer interaction and group decision making. As he stated in his address at the opening of the MIT Center for Collective Intelligence in 2006:

Sometimes collective intelligence is good; sometimes it isn't. Sometimes it works, and sometimes it doesn't. A very important part of our goal is to help put a more solid scientific foundation under the claims in this area.

Fortunately, we don't have to start from scratch in doing that. There's already a lot of good work that has been done in many fields, including psychology, organization theory, artificial intelligence, brain science and others. Part of what we want to do is to help organize the work that has already been done.

But even if we had already organized all of the results of all of the previous research, there would still be a lot to learn. New technologies are now making it possible to organize groups in very new ways, in ways that have never been possible before in the history of humanity. And no one yet understands how to take advantage of these possibilities.[1]

In the chapters that follow, you will see a quilt of technologies, sown together by innovative, passionate people trying to make a difference in their communities. There is no "best" approach. Variations depend mainly upon who is using the early warning information; the levels of political, civic, and religious leaders involved; the speed and accessibility of Internet access; the availability of computers, cell phones, and radios; the technical sophistication of staff members; and the amount of funding available. We can draw from these experiences to learn about the benefits and limits of technology in our attempts to make our troubled world a more peaceful, less violent place.

3 Organizing against Ethnoreligious Violence in Ahmedabad

We were aghast as the minority community hounded out the other few, who did not subscribe to their creed and burnt most houses. We asked ourselves—"what has happened to the community organizations, which were so carefully nurtured over the years . . . ?"

—Cedric Prakash, *Annual Report*

From 1993 to 1997, I had a fellowship with the W. K. Kellogg Foundation that provided funding to investigate how international non-governmental organizations (INGOs) can prevent violence between people of different religious identities. In the winter of 1994, I traveled to Mumbai, where I was told I needed to meet Fr. Cedric Prakash and learn about his work in Ahmedabad, a city in the state of Gujarat in the northwestern part of the country. He was identified as a leader in the field of violence prevention, who worked with people of all faiths, in this religiously cosmopolitan, largest democracy on earth, emerging economic powerhouse called India.

I took a train to Ahmedabad and was startled as I passed through slums on the outskirts of Mumbai and other cities. One often hears about culture shock when traveling in a different place. But this was "poverty shock." I had worked with extremely poor people on the Navajo reservation in New Mexico and in Appalachia in the United States, but I had never seen such destitution, misery, and squalor as I did while traveling on that train.

Prakash was waiting for me at the station when I arrived. Little did I imagine then all that he and his staff members would teach me over a period of three years. They arranged the interviews and accompanied me when I conducted interviews with government officials, civic leaders, religious leaders, staff members of NGOs, members of community-based organizations (CBOs), and slum dwellers.[1]

During my first week in Ahmedabad, we walked through slum after slum, where I was surprised and impressed that Prakash was somewhat of a celebrity. People of all faiths were warm toward him, greeting him with an invitation for "chai" (South Asian tea made by boiling water buffalo milk, tea leaves, and vast quantities of sugar). Clearly, he and his staff members had built substantial goodwill among these impoverished people of many faiths.

What follows is what I learned during those years. Facts and figures reported here are from 1994 to 1996, to accurately reflect the context in which the project I describe was implemented.

The Context

In 1947, British India gained independence from the colonial control of the British Empire.[2] East and West Pakistan, with predominantly Muslim populations, broke off from India, comprised substantially of Hindus. During partition, many people were displaced and killed in identity-based violence along ethnic and religious lines. Since then, there have been numerous disputes and wars between India and Pakistan, the first of which was the secession of West Pakistan from East Pakistan, becoming Bangladesh. Military violence between India and Pakistan continues over the disputed territory of Kashmir, and ethnic tension within India and Pakistan has been fueled, in part, by international tensions.

Identity in Ahmedabad is drawn mainly along ethnoreligious lines. As determined in the 1991 India national census, the overwhelming majority of people, 89 percent, were Hindu. Muslims constituted by far the largest minority at 9 percent. And there were small minorities of Christians, Jains, Sikhs, and Zoroastrians (National Institute of Urban Affairs 1994).

Mahatma Gandhi, the leader of nonviolent resistance to colonial rule, started a protest on March 12, 1930, against the British Salt Law in Ahmedabad. He and his followers objected to the British salt monopoly, contending that Indians should be free to make salt on their own. Ahmedabad became known as an important city in the Indian independence movement and its use of nonviolent resistence.

But this city has had many violent riots between different ethnoreligious groups, mainly in the city's slums, where roughly 41 percent of the total population of 2.9 million lived as of a 1991 census (National Institute of Urban Affairs 1994). Major riots occurred in Ahmedabad during November and December 1990 and in December 1992 (including in Sankalitnagar and Mahajan-no-Vando, two of the slums of particular interest in this chapter) following the destruction of a mosque at Ayodhya by Hindu militants (Patel 1995, 375).

Violence in Ahmedabad during rioting has included stone throwing, hurling burning rags at crowds, destruction of property, police beatings, robbery, and murder. Police have used tear gas regularly to disperse crowds during riots.

It is mostly men who perpetrate the violence during the riots, especially murder. Nevertheless, in an emotional frenzy, women have been seen attacking others, especially other women and children. In addition, women tend to loot, and they, along with children, often serve as propagandists for those cultivating riots.

It is believed generally that rioting in Ahmedabad is cultivated by political leaders, real estate developers, and organized militant groups with foreign support. It also is widely accepted that some leaders of the Bharatiya Janata Party (BJP) have "engineered" riots between Muslims and Hindus as a means of solidifying their political base and of making the Congress (I) party appear weak—consistent with the "greed" dimension of the greed-grievance-nexus analytical construct. Allegedly, BJP operatives have hired "gangsters," some of whom are slum dwellers themselves, to foment identity-based violence along ethnoreligious lines during volatile periods. These periods tend to coincide with and are most predictable during religious holidays, when sectarian collective identity becomes more intensely defined and passions tend to run high, including those which can be readily manipulated with us-versus-them and we-are-on-the-good-side, self-righteous justifications.

In some instances, real estate developers benefit from conflict inasmuch as entire communities—which are living in commercially "developable" locations—are dislocated. They allegedly foment violence on their own or through an "alliance of convenience" with BJP leaders. In either case, there is evidence that suggests that this is a well-founded allegation.[3] In a number of instances, after slum dwellers were dislocated because of a riot, the slums have been "leveled" and turned into "middle-class" neighborhoods.

Information on the role of organized, militant groups is hard to find. According to K. N. Shelat, the district collector based in Ahmedabad, the police are aware of their existence, but counteracting them is difficult.[4]

St. Xavier's Programs

St. Xavier's Social Services Society ("St. Xavier's," for short) began as a humanitarian relief NGO of a Catholic religious order, the Society of Jesus (also known as "the Jesuits"), in the 1970s. Its initial focus was to help those suffering from the flooding of the Sabarmati River in Ahmedabad. Over the years, St. Xavier's developed an extensive array of development programs, too. These are aimed at raising awareness about sanitation and health, providing nonformal education, advocating for human rights (focused mainly on mediation and litigation involving land disputes or spousal abuse), income-generation projects (such as in helping to facilitate the sale of crafts) for women, and promoting environmental stewardship.

St. Xavier's received funding in the early 1990s came from Aide à l'Enfance de l'Inde, Luxembourg; Caritas Neerlandica, Netherlands; Caritas Sverige, Sweden; Catholic Relief Services (CRS), United States; Cebemo, Netherlands; the Government of Basel, Switzerland; Manos Unidas, Spain; Misereor, Germany; and a number of private individuals. Among these international NGOs, CRS is unique in that it provides U.S. government food commodities as well as cash grants. In fact, St. Xavier's, serving in a partner role with CRS, provides food aid to a network of forty-eight operating partners in rural areas throughout the state of Gujarat, involving an extensive food-for-work operation. This food program is important in that the scope of St. Xavier's support for relief and development activities reaches beyond Ahmedabad, as does the administrative authority of prominent government officials with whom St. Xavier's maintains cordial working relations.

St. Xavier's work is largely foreign funded. Occasionally, it receives small grants from the government of India and some private donations from local supporters. In addition, the minor fees charged for medicines in the slums help finance dispensaries.

The annual budget of St. Xavier's, excluding the value of food commodities, averaged roughly $100,000–$125,000 during the 1990s. Of that, close to 75 percent was devoted to St. Xavier's community organizing, training, and other programs in the slums and villages. The rest covered salaries, building rent, and the cost of a documentation center on social, environmental, health, and human rights issues.

The value of food commodities distributed by St. Xavier's was about $1 million annually. Most of this went to the operating partners throughout Gujarat, though St. Xavier's used some commodities during emergencies in Ahmedabad.

St. Xavier's had a professional staff of close to twenty full-time people, with numerous others (especially health workers) working on a part-time contractual basis. The staff members of St. Xavier's included Hindus, Muslims, Jains, and Christians.

Working in the poorest communities in Ahmedabad, St. Xavier's focused on three major, flood-prone slum areas: Sankalitnagar, Mahajan-no-Vando, and Nagori Kabarasthan. The Sankalitnagar slum, where St. Xavier's began working in 1973, had a population of close to twenty-five thousand. Prior to the 1990s, the slum was roughly 60 percent Muslim and 40 percent Hindu. Thereafter, it gradually became almost completely Muslim, with only close to 1 percent Hindu. The Mahajan-no-Vando slum, where St. Xavier's began working in 1983, had a population of close to twelve thousand within a geographical area of 12,807 square meters. It was almost completely Hindu, with only a few Muslim families living on the periphery. Many of the Hindus were dependent on Muslims of the community for petty domestic jobs and on their shops for purchasing goods. The Nagori Kabarasthan slum, where St. Xavier's also began working in 1983, had a population of about eighteen thousand and a geographical area of 10,556 square meters. It was 95 percent Hindu, with a few Muslim families living in the middle and a few more on the periphery.

Beginning in 1992, St. Xavier's began a deliberate outreach program to establish a presence in twenty other slums in Ahmedabad. Most of these outreach areas were sites of St. Xavier's previous emergency activities, when relief tended to be a beginning of an NGO-community relationship. An outreach effort generally began with health and nonformal education services.

The inhabitants of these slums face pressures from a multitude of sources. Most work as manual laborers, domestic helpers, petty retailers, and semi-skilled trade workers. The average monthly family income was $40–$50, which becomes increasingly inadequate due to inflation. The families in these communities face crowding pressures due to population growth and in-migration. The average family dwelling is ten feet by twelve feet. Their homes were built of discarded materials. Four months of the year, when the temperature reached as high as 115 degrees Fahrenheit, the heat inside the tin-roofed homes compeled people to sleep in the filthy, sewage-laden alleyways between the dwellings. It was not uncommon to find ankle-deep water inside them throughout the three-month monsoon season. During high winds, the structures sometimes collapsed, and corrugated metal thrust by the wind caused lacerations.

Relief, Development, and Human Rights Programs

St. Xavier's assisted the government in assessing the number of injuries and damage to property incurred during riots and has provided medical care, food, cooking oil, and blankets to victims of the violence.[5] In some instances, staff members of St. Xavier's found it necessary to search for people who had fled and to provide them transportation back to the site of their former domicile.

As mentioned previously, St. Xavier's relief activities were not confined to the slums in which it has an established presence. Following a riot, St. Xavier's staff went to the affected areas, identified those with genuine riot-related needs, and attempted to provide them with assistance. St. Xavier's did not set up temporary offices there; staff members merely conducted surveys and then distributed resources.

St. Xavier's regularly assisted slum areas affected severely by floods. In such instances, the staff set up health clinics, distributed food, and supplied materials for temporary housing.[6] In addition, St. Xavier's took an active role in a city-wide task force in response to the devastating 1993 flood.

A main focus of St. Xavier's efforts in the slums was on community health. The activities included providing growth monitoring services for children, health education, immunizations, midwifery training, tuberculosis patient treatment, and health outreach when epidemics broke out.[7]

St. Xavier's initially established a nonformal education program, INNovative Education (INNED), in the Sankalitnagar slum. In the early 1990s, it expanded this program to the Mahajan-no-Vando and Nagori Kabarasthan slums. INNED is a supplemental program that encouraged children to participate in formal schooling. Through INNED, children were taken on educational field trips, shown educational films, and given an opportunity to attend educational/leadership camps. The children also performed street plays designed to raise awareness about health and other community issues.

St. Xavier's had a savings program focused specifically on women. It initially did not have a gender differentiation in the program. St. Xavier's found, however, that it was necessary for the women to create separate bank accounts. In so doing, women's savings went up tremendously due to their confidence that the money would be spent in ways agreeable to them.

St. Xavier's cultivated a women's committee in each of the three main slums in which it was working. This committee served as the focal point for the savings scheme as well as a maternal-child health program aimed at prenatal and postnatal care.

Largely through its food-for-work program in rural areas, in collaboration with operating partners, St. Xavier's sought to protect the environment. Program participants engaged in tree planting, watershed management, and wasteland development. In addition, St. Xavier's also promoted the use of alternative energies, especially solar kitchens used in feeding programs.

St. Xavier's had a human rights program designed for the people it served, aimed largely at increasing their awareness of legal protections available to them. This program had a special focus on issues facing women and the prevention of spousal abuse.

While St. Xavier's generally encouraged mediation efforts over litigation, in April 1992 it was instrumental in securing the life sentence of a man for burning his wife to death. This conviction, according to Prakash, markedly reduced spousal abuse.

St. Xavier's also retained legal counsel to protect slum dwellers involved in land disputes. It had more success in preventing eviction than it had in securing titles to land.[8]

Programs to Cultivate Communal Harmony

The conflict prevention and mitigation activities of St. Xavier's took considerable staff creativity to develop. Designing and establishing a program to help prevent violence was, according to Prakash, "not a planned process." It began in 1991, when St. Xavier's staff tried to answer questions about the causes of conflict in the slums. In Prakash's words:

We decided to take a look at the *why* of the whole thing . . . what has to be done to change the situation? We looked at the life of the poor; how they are constantly subjected to many dehumanizing processes. We looked at the role of the building lobby; the land sharks and land brokers who would like some of these poor people out. We looked at the politicians who don't want to come to grips with the situation poor people are in. We concluded that the poor are being used.[9]

From this conclusion, St. Xavier's staff members decided they needed to develop a strategy for a Communal Harmony Project. They understood *communal harmony* to be a state of inter-ethnoreligious relations in which different groups embrace diversity rather than trying to marginalize those who are different. These projects fell into two broad categories: *promotive* and *preemptive*. *Promotive* approaches were designed to cultivate inter-ethnoreligious goodwill. They included integrating communal harmony themes, sponsoring creative competition for children, organizing an annual "people's festival" that encouraged appreciation for or interaction with individuals of other ethnicities, performing symbolic gestures for peace (such as encouraging people of one religious identity to participate in the holy day celebrations of people of another religious identity), and generally integrating the theme into its various relief and development programs. St. Xavier's used various art forms, as well, to inculcate an appreciation for communal harmony.

St. Xavier's sponsored an art competition for slum children on the harmony theme. The idea was to create awareness about the evils of sectarian violence in a fashion that is fun for children. The number of participants increased yearly. Children were provided with drawing paper, watercolor paints, or crayons. In February 1993, roughly fifty children participated.

St. Xavier's cultivated another form of creative competition by sponsoring an essay, poetry, and poster-making contest. Roughly one hundred students from a local Catholic high school participated in 1993.

A similar approach was used by the press. For instance, in December 1993, the Ahmedabad edition of the national English language *Indian Express* newspaper sponsored a competition for excellence in public service advertising. One emphasis of this competition was communal harmony. Assuming a causal relationship between the creative competition of St. Xavier's and that of the press is, perhaps, dubious. Clearly, though, the *Indian Express* editors were aware of St. Xavier's work given that they invited Prakash to be a judge in the newspaper's competition.

During one of the people's festivals, St. Xavier's commissioned a "harmony song." Translated from Gujarati into English, the words are as follows:

Here is the message of communal harmony
Allah and Ishwar are one[10]
Do not fight over a temple or mosque
Politicians fight for power
The huts of the poor are set aflame
The lust for power is the fuel
Look at what has happened to our city
For someone's fault someone else is punished
If, we the people, live in harmony
Nobody will dare to disunite us
This is the message of communal harmony.

St. Xavier's staff members and volunteers distributed the words to this song on plastic sacks that were kept by people because of their intrinsic value, adding to the number of people exposed to the song. It was sung widely throughout the city. This was because, in part, as Prakash put it, the song had "an easy tune like a jingle."

What is fascinating about the song is that the main point is to not let politicians use religion to cause violence. It is an appeal, in very simple but useful terms, to prepare people to view facts unfolding quickly through a lens that allows them to see the greed side of the greed-grievance nexus.

St. Xavier's efforts to use art to "inoculate" people against the mischievous influence of greed-driven manipulation were not limited to music. They also used street plays.[11] One such play is called *Manasjat*. In Gujarati, "manus" literally means "humankind" and "jat" means "caste," as in the Hindu caste system. Hence, "Manasjat" means "the mold of humankind." The translated play follows:

A Scene Full of People
(In a chawl—a slum where people are less poor than in most others—people are gathered, doing their own work. Children are playing.)
(Ranglo and Rangli start singing what amounts to the equivalent of a chorus in a Greek tragedy.)

Ranglo and Rangli: Look . . . look! Our society (community)

Our simple and humble society

Our hard-working society

Where there is unity, there is harmony.

And in harmony there is freedom.

Where more hands get together

There is greater strength.

(After the song a political leader comes into the chawl along with his two chamchas—"goons"—in typical political style.)

Two men: Long live Sir! Long live!
Ranglo: See, see, who is coming.
Rangli: He comes with his two chamchas.
Two men: Long live Sir! Long live! (They go on shouting.)

The Scene of Samjuben's Meeting
(The chawl people shouting—talking to each other, meeting)

One: The "goonda" of the leader had come and ordered us to vacate our homes.
Two: They even offered us money as bribes.
Three: We were born here, our livelihood is here, and we will continue to live here.
Four: And yes, if we are to stay far away then we would have to pay a lot for the to-and-fro bus fare.
Five: We have been living here since several years: how can we leave this place?

Six: They come only for the votes but there is no one to help us in times of difficulty.

Seven: Bad enough that we have to struggle for our own survival; now more pain is inflicted on us.

Samjuben: Dear brothers and sisters, we need to have unity among us. We have to stay together. We have to be one group. And as one body we must fight with them, only then nobody would be able to move us from here. But don't forget, be careful of bribery—don't fall into the trap of greed.

(The meeting continues in mime—the other side, the leader comes in a car. He points out to his Henchman the land that he wants to be vacated.)

Henchmen: We hail you, Sir!

Leader: Salute! Salute! (Angrily) Look here, this land is not merely land but a piece of gold. I therefore want it within two days—understand?

Henchmen: But sir, what about these people?

Leader: Do anything you want. . . . Give money to them. If they don't take it, threaten them and if they still don't agree burn everything in the name of religion and caste. Incite the people that they may hate and kill each other.

Henchmen: It will be done sir it will be done. . . . Don't worry.

(The leader goes away. On the other side the meeting ends. The people talk to each other of "unity" and they go home to sleep.)

(Two goondas come with kerosene and set the chawl on fire.)

Daughter: Mummy, oh, Mummy, wake up! Wake up! There is fire in the house.

Mother: Oh! My God (shouting) HELP! HELP! Please come! Please help! (The daughter starts crying.)

First: We are destroyed!

Second: Oh! My daughter!

Third: Everything is burnt.

(People pour water and calm the fire.)

(Ranglo and Rangli come and sing a song.)

Ranglo and Rangli: From several generations we have been set ablaze.

(Meanwhile the leader comes and isolates the mother and daughter.)

Leader: It is very bad. I am sad. Kindly take this money.

Mother: Uh! We have been looted. Everything we possess has been burnt.

Daughter: (Angrily) Get out of here! First you set us on fire, then you come to console us. Get out! Otherwise something will happen.

(Exit)

(In another corner is the leader with his goondas.)

Leader: (To his men—angrily) Why are these people still here?

Henchmen: Sir, they are not ready to accept any money.

Leader: Then make them fight and kill each other.

(The leader gives money to the goondas and leaves.)

(In two corners: on one side there are Hindu families and on the other side there are Muslim families.)

First Henchman: (To a Hindu) Take this money for your daughter's marriage. You can celebrate it grandly. I am with you but be careful of those Muslims because I heard they are going to attack you.

Second Henchman: (To a Muslim) Why do you look so disappointed?

Muslim: My son is very sick and I don't have money to take him to the hospital.

Second Henchman: (Giving money) Take this money. Our leader has given it to you. I want to tell you one thing. Those Hindu people were planning to attack you all.

The Scene of Rathyatra

(This is the chariot procession in honor of God, the destroyer of life. Also known as Jagannath, it is now called "Juggernaut.")

Muharram (This means "sacred month" and marks the beginning of the Muslim year. This month is associated primarily with a period of mourning. Shiite Muslims observe this month in commemoration of the martyrdom of Ali's son and Muhammad's grandson Hussein. Between the seventh and the tenth day of the Muharram month, processions are held to commemorate this martyrdom.)

Processions:

(Both the Hindu and Muslim crowds get together in their respective areas. They are in an angry mood.)

First Hindu: The big crowd of Muslims seems to be coming towards us.

Second Hindu: Let it come! We will see what happens.

Third Hindu: We have not worn bangles.

Fourth Hindu: Jai Shree Ram!

First Muslim: Hey! They are coming!

Second Muslim: Let them come. Who is frightened of them?

Third Muslim: Ya Allah!

(This time, the henchmen smile at each other and go.)

(From the mobs we can hear the sounds of:)

"Jai Ranchod!" (which means "hail to thee, Ranchood." Ranchod is another name given to Krishna and on the day of his birth people praise him by shouting this slogan.)

"Makhanchor!" (There are several legends or stories about Krishna. One of them is that he constantly stole butter from a pot in which it was kept, so "Makanchor" literally means "butter thief.")

"Badshahi Karbale!" (This is also a slogan shouted on the day of the Muharram procession. Karbale refers to a special place, or ground, that is kept apart for the

Taziahs to be placed or immersed in remembrance of Karbala in the Middle East where Hussein was murdered. This slogan literally means "the King of Karbala.")

"Ya Ali, Ya Ali!" (This is a lamentation remembering Ali, the father of Hussein, who was also martyred at the same time.) (Both of the mobs come in front and stand.)

First Hindu: For years our Raths (chariot processions) have been passing through this way only.

Second Hindu: Our Ranchhodji (This is a respectful form of addressing Lord Krishna) will pass through this way only, today.

Third Hindu: Whatever may happen!

First Muslim: From centuries our Tajiyas have been passing through this way.

Second Muslim: Today it will pass through this way only.

Third Muslim: Ya Ali! Both the mobs fight each other—kill each other, injure each other.

(Sounds of shouting, crying, and yelling)

Ranglo and Rangli: (Singing)

One says it is my right

The other says it is mine

Humans say let us swallow each other

Hindu woman: But why?

Muslim Woman: But for what?

Ranglo and Rangli: (Singing)

For peace and for selfish gain

To grab and to control

They learn and they teach

They sing and they make them sing

They see and they show

That is the nature of humankind.

Hindu woman: But how?

Ranglo and Rangli: (Singing)

By instigating and inflaming passions

By dividing and ruling

By dividing and ruling

(After the song)

(Hindus and Muslims go to their relief camp and sit. Meanwhile a social worker, Samjuben, comes to distribute items and food. She consoles the people.)

First: We are looted!

Second: We are killed!

Third: Oh God! What happened?

Fourth: Now where will we go?

Fifth: We are on the footpath now.

Sixth: I don't want your blanket or oil. I want my brother. I want my brother!

(Crying)

(Meanwhile, the leader comes in a motorcar. He consoles everybody.)

Ranglo and Rangli: (Singing)

Here comes a total hypocrite

All dressed in political garb . . .

Leader: I am very sad with what happened. I cannot bear to see your state. Our country, India, is a democratic one. And in a democracy, we are all free to practice any religion. The Father of our Nation, Mahatma Gandhiji, said "Hindus and Muslims are like my two eyes I promise that all who have been injured and looted will get five thousand rupees from my funds." (The chamchas tell others to clap.)

(This time some people in the relief camp show their anger toward the leader. They throw the ribbon they were given, identifying them as Hindu or Muslim. They hit the leader and say:)

All: Netaji Murdabad! Netaji Murdabad! (This is a highly insulting slogan that means "Down with the "Politician!")

Ranglo and Rangli: We were one. We will always remain one. Manasjat . . . Manasjat!

All: (Singing) We shall overcome. . . . Manasjat!

As with the song St. Xavier's commissioned, the play helps people see the folly of allowing themselves to be manipulated into violence. The Manasjat play, among others, drew large crowds in the slums. Some of the plays also have the effect of fostering emotional catharsis. A central theme is that the people who are hurt most by riots are the slum dwellers themselves. Street plays are used when tension is increasing, after violent uprisings, or as a means of keeping the importance of communal harmony in front of people.

Two underlying traits of the plays are relevancy and simplicity. Each play is tailored to a local situation. In the view of the staff of St. Xavier's, a play must use symbols and words that are common to the people who will see the play. The play's language must be changed regularly, incorporating updated information and new symbols. St. Xavier's staff concur that professional playwrights would likely develop a play ignoring local idiosyncrasies; for street plays to be effective, they argue, the writers must be local, and have intimate knowledge of the community.

The initial Manasjat street play was developed during a staff workshop in 1991, with a consultant providing guidance on the processes whereby such art forms are created. The staff, however, wrote the play. Staff members divided themselves into groups and presented their versions. Finally, a script was written by borrowing the best ideas from each of the various presentations. Since then, in subsequent pieces, no street play trainer/consultant was used. In explaining why, Prakash commented that the consultant "didn't know the slums too well."

According to staff of St. Xavier's, for a street play to be effective it must be "organic." Such plays, they argued, are even more appealing to slum dwellers than television. Despite their poverty, television viewing was somewhat common among slum dwellers. But the television programs were, the staff members point out, "far away. . . . [In contrast to] street plays, we are using their idioms, their folklore, their customs, their phrases, their words, their myths. They are seeing themselves in the play. The play is real, not artificial."

Another trait, simplicity, is important because as Prakash put it: "The whole idea of a street play is that people are able to imbibe the meaning with simple phrases, simple verses, simple rhythms."

In addition to these *promotive* approaches, St. Xavier's Communal Harmony Project used the following three *preemptive* approaches in its violence prevention approach:

1. Providing a Safe Haven
One of the earliest approaches St. Xavier's used to prevent violence was to intervene prior to imminent hostilities in a disturbed community and provide safe haven for a besieged minority.[12] This tactic requires considerable skill in mediation to convince the aggressive party to not pursue potential victims. In this capacity, Prakash felt that he had an advantage as a Christian because his religion enhanced his ability to be perceived as an objective third party during Muslim-Hindu confrontations. However, other, non-Christian staff members of St. Xavier's and slum dwellers themselves also played this role.

It should be noted that the mediation skill of slum dwellers was cultivated deliberately by St. Xavier's using role playing during various workshops (those in which health workers, for instance, were trained). These activities had no formal structure, nor was there a pre-set script or guidebook. Instead, the workshop facilitator used improvisation and asked the participants to develop their own role-play exercise.

According to Prakash, St. Xavier's cultivation of indigenous mediation was aimed mainly at modifying the "atmosphere" surrounding peoples' response to conflict, changing their "mindset," and cultivating their "confidence and courage to be able to do this."

2. "Myth Busting"

Propagation of myths is, both figuratively and literally, a fine art in some of Ahmedabad's slums. During particularly volatile periods, usually before a religious festival, political operatives (strangers to the area) paid slum children a rupee or two to hand out inflammatory *patrikas* ("brochures" or "handbills"), designed to foment a violent uprising. These patrikas typically referred to offensive behavior of people in a different community. For instance, when Pakistan beat India in cricket, patrikas were sometimes distributed which said that the Muslims had been cheering for Pakistan. Even though most of the people in the slums were illiterate, the novelty of receiving a piece of paper with a message for them led them to ask a literate member of their community to read the patrika out loud.

The political operatives also hung posters and painted slogans on walls as a means of communicating. The inflammatory messages often referred to "the other community" and what they have "done to our country," asking people not to patronize "their" shops (Prakash 1994a, 8).

Once when such a cricket match myth was propagated, St. Xavier's staff went into the communities and asked: "Have you seen any Muslims cheering?" (i.e., using a strategy of refutation by counterexample) and "If it is a good stroke [i.e., "hit"] why can't you applaud it?" (i.e., appealing to collegial sentiments to be "good sports"). The essence of this strategy, according to Prakash, is "to counter false propaganda as soon as it takes off—bit by bit and point by point" (1994a, 12). Generally speaking, this was done by holding community meetings called by peace committees.

It should be noted that myth busting occurs at an official level as well, with the media. K. N. Shelat, the district collector, regularly briefed the press, especially when interfaith tensions were high.

3. Appealing to Authorities

A final preemptive approach is in appealing directly for the police to intervene. Of course, in instances in which police officers are partial to a certain group, their intervention can exacerbate, rather than mitigate, violence. In such cases, St. Xavier's typically appealed to moderate mid- and top-level governmental leaders who would instruct subordinates to intervene. Sometimes these leaders appealed for an intervention by the military. Some examples follow.

Evidence of Success

I found three instances when St. Xavier's staff members were instrumental in preventing violence. First, on December 14, 1990, residents of the Miriyam-Bibi chawl of Gomtipur were attacked by police and militant Hindus from outside the area. Five thousand Muslims took refuge in St. Mary's Nursing Home during the middle of the night, after the police and a Hindu mob broke into their homes, raped some of their wives and murdered others. Essentially, they made their own assessment about the danger of impending violence using inductive reasoning. The nuns who ran the nursing home called Prakash since they knew of St. Xavier's work in communal harmony. He called the district collector, who invoked his powers as executive magistrate, ordering the police to retreat and replacing them with the Army. The Army disbanded the mob of Hindus.

The second instance was during the winter of 1991–1992 in Shahpur (one of the "outreach" slums), where a group of militant Hindus threatened to attack Muslims. Hindus living in the slum took it upon themselves to "get in the way" of the militants, telling them "you kill us first."[13] Though it mitigated violence, the actions of these Hindu slum dwellers occurred just prior to an attack, suggesting that neighborhood residents heard of it just before it was about to happen. The intervention was "organic"; it was neither organized by peace committee members nor by the staff members of St. Xavier's. It is difficult to determine whether this spontaneous behavior on the part of Hindus in the slum was a result of prior activities to promote communal harmony, or whether it occurred for other reasons (such as close neighborly ties and enlightened leadership). But violence prevention occurred very closely to when violence would have erupted otherwise.

The third instance occurred in January 1993, when middle-class Muslims living on the edge of Mahajan-no-Vando planned to attack Hindu slum dwellers following the destruction of a mosque in Ayodhya (which was symbolically potent throughout India). In this case, violence was prevented when Prakash contacted Muslim civic leaders urging that they confront the Muslim militants with a plea for restraint. The civic leaders did so and were successful in preventing violence. As such, this intervention is an example of the potential value of communication with mid-level leaders holding positions of influence to prevent violence, as was also evidenced by the intervention of the district collector mentioned earlier.

Evidence of communal harmony, which staff members of St. Xavier's felt reflected their work, and which may or may not have resulted in reduced communal violence, includes the following:

• *Rakhis* (ceremonial bracelets signifying a protective sister-brother relationship)[14] were given by Hindu women to Muslim men in the Mahajan-no-Vando community in 1993. This community had experienced communal violence leading up to a flag-hoisting ceremony designed to promote communal harmony earlier that same year. In fact, at the ceremony, the tension was so high that staff members of St. Xavier's stood back in the anonymity of the crowd for fear of their own physical safety. The exchange of *rakhis*, therefore, signified a significant thaw in interfaith relations.

• The street plays evoked emotional reactions. Viewers often left the plays in tears. In one instance, according to one St. Xavier's staff member, a woman "went into hysterics." Staff members argued that the plays were not intended to be recreational or entertaining. They were designed to help people face the truth, to make them think, to help them overcome petty jealousies and hatred. To the extent that catharsis is important to bring about forgiveness, these emotional reactions arguably are indicative of the plays' positive impact on communal relations.

• Majority Hindus helped minority Muslims when the government imposed a curfew in anticipation of a riot. People of both faiths cross religious lines to share food and water, assist victims of looting, and take the injured to the hospital.

• Muslims are invited to Hindu festivals in Ahmedabad, and Hindus are invited to Muslim festivals.

• During a volatile time, a Hindu man was heard saying to a Muslim, "You are like my son. Whenever you feel threatened, you come to my house."

• Hindus were seen sitting on the doorstep of Muslim dwellings when Hindu communal passions are running high.

• In one slum, after stones were thrown maliciously, both Hindus and Muslims went to see who was throwing them rather than assuming it was someone of the other religion. They found that the perpetrator came from outside the community, thereby averting misunderstanding among slum dwellers.

• Not only was the "harmony song" sung throughout the city, but also the harmony theme was embraced by other institutions of influence, namely the press. The fact that a major daily newspaper in Ahmedabad sponsored a harmony-related advertising competition suggests that a social marketing approach of this type can have an *agenda-setting function*, meaning that it has an impact on the focus of attention within the community (McCombs and Shaw 1972).

• There is an element of agency self-flattery that suggests success in using street plays. Whereas they were started initially as a way to prevent violence or to help people with their emotions after a violent confrontation, St. Xavier's subsequently expanded the use of street plays to new subjects. These included problems related to liquor consumption, suicide, and poor hygiene. The staff members' enthusiasm and broad use of the approach suggest they at least consider the conflict-related plays to be worthwhile.

Signs of Failure

But there were instances when St. Xavier's approach was unsuccessful, as well. Based on his experience, Prakash maintained that St. Xavier's approach was successful in overcoming tension resulting from social or political pressures and cajoling. But he argued that, despite all of the myth busting, trust building, and community organizing that was done, religious symbols, when manipulated in certain ways, still seemed to invoke an undesirable response in the slum areas in which St. Xavier's worked.

There are two instances in which the strategy failed to prevent religious symbol-related violence. First, in December 1992, mob violence erupted in the Sankalitnagar and Mahajan-no-Vando slums (the latter of which had two Hindu youth who participated in the destruction of the Ayodhya Mosque). During this period, staff members of St. Xavier's were "hounded" out of the slums, despite St. Xavier's years of extensive involvement in both communities. In commenting on the violence in Sankalitnagar, Prakash wrote that it "shook the very foundations of years of innovative and pioneering work done by . . . [St. Xavier's], in community development and in low-cost housing. . . . We were aghast as the minority community (which constituted a majority there) hounded out the other few, who did not subscribe to their creed and burnt most houses. We asked ourselves— 'what has happened to the community organizations, which were so carefully nurtured over the years?' Somewhere, something seriously had gone wrong." Indeed, communal hostility in the slums apparently spilled over into hostility toward other programs of St. Xavier's. Commenting on this, Prakash noted that in Mahajan-no-Vando "the people were apparently hostile to the INNED [education] classes being run there and the closest they came to cooperation was non-interference. We constantly had to run from pillar to post trying to get the people [to] make available a room or a shed to continue educational classes. We have not been successful in getting the parents/guardians to run the Nutrition Improvement Program in the area."[15]

The second time the strategy did not work occurred in July 1993, when a tiny shrine on a main road of Ahmedabad near the Shahpur Fire Brigade Station was turned into a fairly large shrine overnight. Despite protests from the fire brigade officials that such a shrine could not stay on public property, the people of Nagori Kabarasthan were adamant that it not be torn down. People with whom St. Xavier's had worked for years were heard saying that they were willing to sacrifice their lives for the newly built shrine. Tensions ran high, but the local police were able to intervene and prevent a riot.[16] In the end, the police used their authority in threatening incarceration to increase the risk of collective violence (counteracting the sense of reduced risk as identified by Horowitz). St. Xavier's staff members considered the incident to be evidence of program failure since they had worked in the community for eight years with their Communal Harmony Project.[17]

Analysis

The Problem of Religious Symbols

Perhaps it is correct that this approach—as effective as it may be in using rational arguments to counter strong emotions—is only partially effective (helpful, but not sufficient) when religious symbols are involved because, as Prakash put it, "emotion overtakes reason and fundamentalism exacerbates this tendency."[18] One must differentiate religious *identities* from religious *symbols*. On the one hand, there is violence between groups with different religious identities about nonreligious issues. On the other, there is violence between groups with differing religious identities in which religious symbols are involved. The strategies employed by St. Xavier's seem to be successful in the former case, but not completely in the latter case. So when propaganda about a cricket match between India and Pakistan encourages a riot between Hindus and Muslims, the foundation laid through this approach, along with a "myth-busting" scheme, tends to be successful in preventing violence, provided that the foundation is strong and the myth busting is timely.

However, when actual or potential destruction of a shrine is involved, the strategy does not work as well, despite a firm foundation and explicit efforts to prevent an eruption of violence in a timely manner, save when it is combined with an effective policing action by the government. In short, the approach seems to work on its own to prevent violence among groups with differing religious identities when the issue is of secular, not religious, significance. It may be partially effective, as one among other approaches, in a situation involving religious symbols. It succeeds when impartial police or military officials intervene (again, consistent with the view that the sense of risk of becoming violent must be increased).

Perhaps this partial failure suggests that a focus on religious leaders—muftis, swamis, and the like—should be incorporated into the strategy, to the extent that those leaders can address effectively an attack on religious symbols which that community holds dear. Even though St. Xavier's cultivates community leadership through a committee system, it does not involve religious leaders in any systematic fashion, due, in part, according to Prakash, to the tendency for many religious leaders to exacerbate rather than seek to counteract xenophobic religious nationalism. He admits, however, that a selected group of religious leaders do help prevent violence. Therefore, involving religious leaders for the sake of communal harmony might be constructive when done selectively.

The Role of Indigenous and International NGOs

It is unlikely that a newcomer NGO, especially an international NGO, would have the credibility and trust of the various communities to be able to develop a communal harmony approach to violence prevention. This strategy, therefore, is suitable for national NGOs with established track records in relief and development, which have credibility with and the trust of the communities in which they work.

Some would argue that an "arms-length" relationship between an international NGO and a local partner using a communal harmony approach is probably advisable, especially when used to avert the effects of manipulation by a political party. Clearly, a foreign agency's support for activities to counter the efforts of a political party could raise the specter of foreign meddling in local or national affairs. Whether the international NGO directly funds the communal harmony activities of the indigenous NGO or only supports relief and development activities, an accusation of foreign meddling can, of course, be leveled.

Staff Security

Staff members of NGOs who get involved in a project to foster communal harmony may, at times, risk being subjected to physical violence. According to Prakash: "When St. Xavier's staff members say that you are being manipulated and everyone knows it is a certain political party doing it, it can be a problem. That party takes offense. That is one of the risks we have to take. Anything like this is basically confrontationist."

It would be a stretch to say that staff members of St. Xavier's are unanimous in their enthusiasm for the communal harmony strategy. Once, during a staff discussion, Prakash asked staff members if they were willing to live with the risks such a strategy brings. They said nothing, but they smiled; some smiled nervously. Indeed, as discussed previously, some of the staff chose to stay back, in the anonymity of the crowd, during the flag-hoisting ceremony designed to promote communal harmony following the riots of December 1992 to January 1993. Reflecting on it later, Prakash wrote that the staff had a "general fear of taking a stand for what is right and running the risk of 'personal equations' changing . . . during the ceremony, some of our staff were hardly involved—preferring to remain mere spectators—rather than take or encourage the people to take a responsible and positive role."[19]

It should be noted, however, that staff members did not encounter threats leveled against them as individuals (such as an intimidating phone call at home). According to the staff of St. Xavier's, security is enhanced by staying away from large crowds whenever possible; not trying to mediate a large mob when there are too few police around for protection;[20] and not asking people who have been recently dislocated by violence to go home (when they are filled with a high degree of insecurity).

Spreading the Use of the Approach

In the summer of 1994, forty-six participants in India from Maharashtra, Goa, and Gujarat met for a three-day workshop on *communalism* (a label used in South Asia for what is often called religious nationalism, which combines religious passion with political ambition parochially, as if one ethnoreligious group is superior to others). The workshop participants concluded that each organization should designate one person exclusively for violence prevention activities. The workshop, which convened NGO representatives, journalists, and academics, was organized by the Center for Social Studies in South Gujarat. According to Prakash, this type of networking activity specifically on communalism is rare. When there are such gatherings, the focus, he argues, tends to be on "what to do about conflict" rather than on "how to cultivate harmony."

The Question of Abandoning Religiosity

There is a fundamental question of whether a strategy like this can or should "water down" the religious views of the various groups, or should celebrate the diversity of those views instead. Prakash argues that St. Xavier's is attempting to get Muslims to be good Muslims and Hindus to be good Hindus. Of course, being a good, devout religious person is a matter of interpretation, but clearly one can highlight the faith components which venerate peace over violence. One would think that it would be more effective to help those with strong religious passions to channel those passions rather than to try to water them down.

The distinction here is between a melting pot of religions, resulting in an indistinct gray matter, versus a spicy stew of them, in which the ingredients maintain their original shape and texture, but their combined flavor is better than what they taste like separately. One downplays religious diversity; the other celebrates it.

The question is whether a strategy that downplays religious differences is as effective as one that celebrates religious differences, especially when it is aimed at counteracting an influence which highlights religious differences as a means of fomenting prejudice and hatred. Does a downplaying approach, in a sense, ask people to abandon their firmly held religious beliefs, putting them into a sort of double-bind of: "Do I maintain those beliefs which I hold dear and fight those who don't hold them, or do I abandon the beliefs I hold dear and not fight?" Is it not less cognitively dissonant to maintain the strongly held beliefs but to be helped to see them as something to celebrate in respectful coexistence?

Problems and Solutions at the Same Level

This case raises a "levels of intervention" question that suggests that when violence is being fomented at a local level, at least a part of the solution must be at the same level. That is, there is a problem-solution parity principle at play. For instance, when violence is being fueled by what amounts to an international organized crime syndicate, then part of the solution needs to be at that level as well.[21] Or, conversely, when violence is being cultivated at a local level, at least part of the response to it needs to be at that level.

Summary

The Communal Harmony Project of St. Xavier's in Ahmedabad is an example of a violence prevention approach that did not use a computerized events data system. Cell phones were few and far between. The Internet did not play a role. The approach used community organizing to build local capacity to detect incipient violence and to respond to it quickly by involving members of the affected communities and by communicating with those of authority with a plea for intervention. The strategy is both horizontal and vertical in the levels of leaders it engages.

The results in preventing violence were mixed, but promising. St. Xavier's successfully prevented bloodshed when providing safe haven for those about to be attacked, by cultivating an appreciation for communal harmony, and by appealing to moderates at local- and mid-levels with a plea for intervention to prevent an attack.

The appreciation for communal harmony that was cultivated through "promotive approaches"—art contests, songs, and street plays—arguably extended far into the community at large, not just among peace committee members. People of the same identity as an attacking group "got in the way" of attackers. Those who stood up for peace had gotten the message. They understood it well enough to think on their feet when tensions and violent passions were high. The "promotive approaches" served as an "inoculation" of sorts against manipulation of passions that were typically cultivated by operatives of those who had something to gain from violence.

But in cases when efforts to prevent violence were unsuccessful, religious passion was substantial and, in one instance, a show of force by police officers was necessary to prevent bloodshed. The presence of police enhanced a sense of risk for violence, effectively short circuiting consensus building (as Horowitz's framework about risk would portend) to insist that a temple be constructed on public land.

It is difficult to draw conclusions about the timeliness of warnings from St. Xavier's experience. It is clear, however, that warnings were a function of inductive reasoning that benefited from the web of relationships developed by St. Xavier's at various leadership levels within the community.

St. Xavier's did not receive encouragement from any of its foreign funders to pursue a violence prevention program. (All of them, however, have provided moral support since the violence prevention work began.) St. Xavier's is an example of an indigenous NGO that did not need early warning experts from the outside to provide guidance on the importance of being diligent about looking for signs of likely violence. Its Communal Harmony Project was developed "organically" out of a felt need articulated by staff and community leaders. Its successes and failures are a helpful baseline against which to compare other programs that use more technology, some of which is provided with support from outsiders.

4   Interrupting Gang Violence in Chicago

Punishment is not part of the game; punishment is completely overvalued; punishment is not how we learn behaviors. It's a very, very, very small part of the whole thing. Behaviors are learned by copying and imitating, observing, modeling, doing what you think you're supposed to do just by watching what other people do. People aren't even caring about punishment half the time. What they learn is how to avoid the punishment. They don't learn that that's what they're supposed [to do]—'cause they're more concerned about what their friends think and what their older brothers think than this other stuff.

—Gary Slutkin, Executive Director, The Chicago Project for Violence Prevention[1]

If cleaning up sewer systems could prevent more deaths than all the physicians in the world, then perhaps reforming the social, economic, and legal institutions that systematically humiliate people can do more to prevent violence than all the preaching and punishing in the world. The task before us now is to integrate the psychodynamic understanding of shame and guilt and the broader social and economic factors that intensify those feelings to murderous and suicidal extremes on a mass scale.

—James Gilligan, *Violence: Our Deadly Epidemic and Its Causes*

Perhaps the best example of training and preparedness for violence interruption is the CeaseFire program in Chicago. This program has directly prevented hundreds upon hundreds of killings in the past few years alone.

—Patrick Meier, "Early Warning Systems and the Prevention of Violent Conflict"

I participate in a working group on campus looking into potential educational uses of virtual reality. Our primary focus is Second Life. This web-based social media system allows users to develop avatars of themselves and to interact with others. Avatars meet on "islands" that are designed to reflect different settings.[2]

I decided to see if anyone was using Second Life to train people in violence prevention. I stumbled across a website for the Center for the Advancement of Distance Education (CADE) of the University of Illinois at Chicago, School of Public Health.[3] The virtual reality site they developed, modeled after an African American neighborhood, is called "CeaseFire Island Life in the Hood." A second island, CeaseFire Isla, "is largely Latino in look and feel" according to Candice Kane, CeaseFire's chief operating officer.[4]

CeaseFire has used these sites to train people how to intervene to prevent violence. Trainees use their avatars during role-playing exercises. Participants move their avatars using a mouse.

I became intrigued not only in the Second Life site, but also by CeaseFire's overall approach. I was pleased to see a program in the United States that is similar to what I saw in Ahmedabad.

This chapter covers the major components of CeaseFire and this program's results. It also addresses the use of a creative Short Message Service (SMS) text messaging system and examines how well the Second Life site has worked for training.

Background

Legendary organized crime in Chicago invokes images of Al Capone, bootlegging, and mafia-style shootouts. Some might think of the 2009 movie *Public Enemies* about the life of gangster John Dillinger.

According to the National Gang Intelligence Center (2009), there are approximately one million gang members engaged primarily in drug trafficking in the United States, up from 800,000 members in 2005. In Chicago, modern gang violence has been endemic and more difficult to reduce than in many other large cities. In 2008, there were 229 gang-related murders in Chicago, constituting 45 percent of the city's total. In 2009, there were 158—34 percent of which were murders. The Organized Crime Unit of the Chicago Police Department, which specializes in cases involving narcotics and gangs, made 8,035 arrests in 2009, confiscating drugs with an estimated value of $208 million (Chicago Police Department 2010, 29, 53).

A major reason why the gang problem in Chicago has been so intractable is because of how the city has handled its public housing. According to Guarino, "gang warfare intensified in Chicago over the past 15 years as a result of the dismantling of public housing projects. . . . Residents were scattered across the edges of the city, opening new turf battles as gangs fought to control drug markets" (2010). As we saw in Ahmedabad, struggles over territorial control are often marked with violence.

The Chicago Project for Violence Prevention was initiated in 1995 as a public health program, employing techniques for violence prevention similar to those used to eradicate infectious diseases—as explained in chapter 1. CeaseFire was created by the Chicago Project in 2000. It was launched initially in West Garfield Park, one of the city's "hot spots."

The program involves five main components. The first is *community mobilization* designed to change attitudes and behaviors toward violence. This entails cultivating and supporting neighborhood coalitions including youth organizations, police officers, religious leaders, block clubs, and individual residents. Their members are asked to distribute public education materials, to identify the extent of violence, and to set goals toward stopping it.

The second component is *youth outreach*. CeaseFire typically engages former gang members as outreach workers due mainly to their "street smarts" and their knowledge of youth in a given neighborhood. Outreach workers are trained in ways to build trust with high-risk individuals and to help them identify and pursue alternatives, such as returning to school, getting a job, or disengaging from a gang. Some of these outreach workers are trained specifically to mediate between gangs during situations of acute tension. These "violence interrupters" have typically served time in prison and can speak about the risks of gang-related violence from first-hand experience. Their approach to early warning is to be in the neighborhoods where gang violence is likely, so they hear about a potential violent confrontation ahead of time.

Recruiting and retaining violence interrupters is difficult. According to Nancy Ritter, a member of the team that evaluated CeaseFire under the auspices of the U.S. Department of Justice:

Violence interrupters must work in the netherworld of street gangs and pass muster with gang leaders. Interrupters cruise the streets of the toughest neighborhoods to identify and intervene in gang-related conflicts before they intensify. If a shooting has occurred, they seek out the victim's friends and relatives and try to prevent a retaliatory shooting.

People who fit this bill often lack traditional workplace experience, and finding and hiring them is not easy. . . . However, the violence interrupters interviewed for the evaluation said they had turned their lives around and wanted to help others do the same.[5]

CeaseFire has safeguards to ensure that the violence interrupters stay clean—that they are not abusing drugs or using violence themselves. They are hired by a panel that includes police officers and local leaders.

Background checks are run, with particular attention given to crimes against women and children. Some sites do not hire anyone with a felony conviction. Violence interrupters have to pass periodic drug tests.

Public education is the third component of the Chicago Project. CeaseFire provides bumper stickers, T-shirts, posters, buttons, leaflets, and other materials to people in the neighborhood with messages designed to raise awareness about the consequences of violence. These materials are distributed and posted by mobilized community members. Here are examples of their messages:

Don't let 6 × 9 [on a picture of a jail cell] or 6 feet under [on a picture of a grave stone] Be Your Only Choices. Don't shoot! Get Help Now! Toll Free: 866-TO-CEASE.

Is your child in a gang? Ask him before it is too late [with a picture of a father crying over his dead son].

Don't shoot. I want to grow up [with a picture of a young boy].

The fourth component: *engaging religious leaders* in violence prevention campaigns more intensively than other members of the community. Cease-Fire staff members ask religious leaders to undertake activities that complement the work of outreach workers. These leaders provide guidance to high-risk individuals, urge their congregations to become actively involved in violence prevention and, in extreme circumstances, they become involved directly when tensions run high.[6] Similar to St. Xavier's, they also provide safe haven for those in danger.

In 1999, 170 religious leaders in Chicago signed the Covenant for Peace in Action that reads:

Proclaim that we will tolerate **NO MORE SHOOTING.**

Declare peace the norm and make it so.

Preach for peace and against violence from our pulpits the first weekend of every month.

Exhort our congregations to work for peace.

Pray and speak for peace in our congregations and on the street.

Assert a strong presence on the streets in response to every shooting in designated CeaseFire zones.

Counsel and support those who seek to change their lives through the provision of positive alternatives.

Adopt, mentor, and open safe havens for the youth in our communities.

Actively rally and work against guns, gun use, and gun trafficking.

Actively work with each other in the neighborhood and in this Task Force to continue the purposes and momentum of this day.[7]

And, the fifth and final component of the Chicago Project, CeaseFire facilitates the participation of the *justice system*, involving police officers, juvenile officers, the courts, and corrections agencies. Officials of the justice system are asked to notify CeaseFire when there are indications that violence is likely or has occurred in neighborhoods where projects are being implemented so that violence interrupters and others can intervene to prevent escalation of violence.

In 2005, CeaseFire developed a program tailored to hospitals. When gang-related violence occurs, victims in emergency rooms, accompanied by fellow gang members, often serve as a magnet for retaliation killings by the opposing gang. Initially hospital staff members at Christ Advocate Medical Center in Oak Lawn, Illinois, were trained in how to handle such cases so that retribution-related violence is minimized or eliminated. The hospital provides funding for two full-time CeaseFire hospital workers. During its initial two-year period, these hospital responders worked with 640 gunshot victims and 55 stab-wound victims.[8]

In 2007, CeaseFire received funding from the Robert Woods Johnson Foundation to replicate their approach in other cities, starting initially in Baltimore, Maryland, and Kansas City, Missouri. Additional cities have established affiliate programs, or are in the process of doing so, in New York, Arizona, Ohio, Louisiana, and Pennsylvania. A taskforce in South Bend, Indiana, is working toward a modified version of CeaseFire's approach by developing a pilot project at Memorial Hospital and engaging the SchoolTipline, a resource for schools through which students send web-based or text messages to a central location to notify school officials of potential violence and inappropriate behavior (such as bullying).[9]

In addition, CeaseFire affiliate projects have been set up internationally. The American Islamic Congress was CeaseFire's first international replication partner, launching programs in Iraq. Other locations include Trinidad and Tobago, England, and South Africa. In addition, other affiliates are being developed in Mexico, Jamaica, Brazil, Democratic Republic of Congo, Sudan, England, and Kyrgyzstan.

Theory of Change and Results

As explained earlier, Wesley G. Skogan, Susan M. Hartnett, Natalie Bump, and Jill Dubois of Northwestern University conducted an in-depth evaluation of CeaseFire, with the support of the U.S. Department of Justice (Skogan et al. 2008). It involved in-depth interviews, analyses of "hot spot" maps, assessment of the program's impact on gang networks, and quantitative analyses of indicators of violence. The evaluation does not cover CeaseFire's international initiatives.

The evaluators were struck by their finding that CeaseFire has a coherent program theory of change. Specifically, as explained in chapter 1, this theory is based on three levers (a common terminology in public health circles) that can be pulled to impact change: widening the decision alternatives, changing norms, and impacting the sense of risk of becoming violent. The program inputs include *street intervention, client outreach, clergy involvement, community mobilization,* an *educational campaign,* and *involvement of police and the courts for prosecution.* The inputs and levers are designed to influence the output goal—a *reduction of violence.*[10] A helpful summary of the approach is provided by Skogan and his team:

The program aimed at changing operative norms regarding violence, both in the wider community and among its clients. Norms are the beliefs, attitudes and values that make up the culture of a community and define the range of behavior that is normally acceptable. Community mobilization, a public education campaign and the mentoring efforts of outreach workers were calculated to influence beliefs about the appropriateness of violence. A second goal was to provide on-the-spot alternatives to violence when gangs and individuals on the street were making behavior decisions. CeaseFire treated the young men and women they encountered as rational actors, capable of making choices. The strategy was to promote their consideration of a broader array of responses to situations that too frequently elicited shootings and killings as a problem solving tactic. This reflected the often accurate view that a great deal of street violence is surprisingly casual in character; people shoot one another in response to perceived slights to their character or reputation, in disputes over women, or for driving through the wrong neighborhood. Worse, in the gang world, one shooting frequently leads to another, perpetuating a cycle of violence. Once initiated, retaliatory violence can send neighborhoods down a spiral of tit-for-tat killings. Finally, the program aimed at increasing the perceived risks and costs of involvement in violence among high-risk (largely) young people. The risk component reflects a classic deterrence model of human behavior, for among the risks that are highlighted are incarceration, injury, and death. In addition, staff members emphasized the "social risks" of involvement, including the impact of violence on the families of clients and the immediate community. The risk component of the model led to a strategic decision to largely hire staff members who could gain the attention of target audiences and communicate these messages credibly. (Skogan et al. 2008, ES-1-2)

The evaluation found that in four Chicago locations "the introduction of CeaseFire was associated with distinct and statistically significant declines in the broadest measure of actual and attempted shootings, declines that ranged from 17 to 24 percent. In four partially overlapping sites there were distinctive declines in the number of persons actually shot ranging from 16 to 34 percent" (Skogan et al 2008, ES-18).

Other findings included: retaliation murders were reduced 100 percent in five of eight neighborhoods; 85 to 99 percent of high-risk clients who needed help (such as in finding a job) received that help; 99 percent of at-risk youth reported that CeaseFire had a positive effect on their lives; and neighborhood safety improved in every community studied.[11]

One of the important features of CeaseFire's approach is its focus on a small number of people, namely those "being shot or being a shooter." Evaluators found that violence interrupters often operated in pairs and are focused on a dozen or fewer at-risk young people at a time (Skogan et al. 2008, ES-1). A profile of CeaseFire's clientele is provided in figure 4.1, and their numbers of arrests are in figure 4.2.

At-risk youth who were interviewed emphasized the importance of outreach workers, who provided assistance in a host of areas, including finding a place to live and getting a job. Data on types and amounts of assistance are provided in figure 4.3. Researchers found that these relationships were most important during critical times, when at-risk youth "were tempted to resume taking drugs, were involved in illegal activities, or when they felt that violence was imminent."[12]

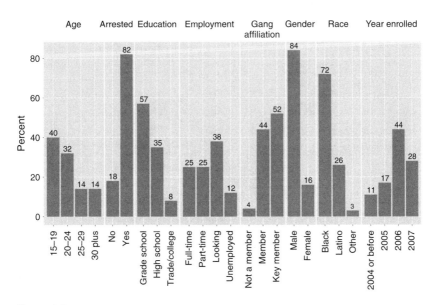

Figure 4.1
Profile of clientele. Permission to use this figure was provided by Wesley G. Skogan.

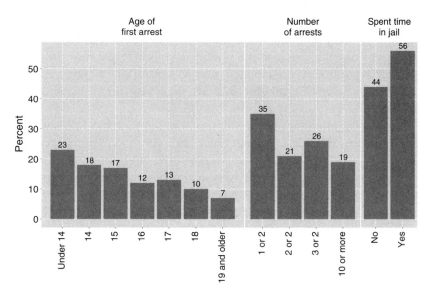

Figure 4.2
Arrests. Permission to use this figure was provided by Wesley G. Skogan.

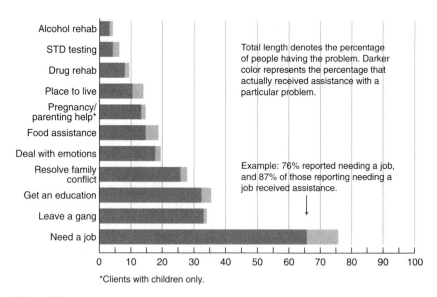

Figure 4.3
Assistance provided. Permission to use this figure was provided by Wesley G. Skogan.

Use of Technology

CeaseFire managers see numerous benefits in using Second Life for training. First, it is a way to make training more available in other locations.
Training programs can be conducted in other cities, even other countries,
by CeaseFire staff members in Chicago. Another advantage is that trainers
witness the performance of trainees and can give them immediate feedback. Coaching can reinforce desirable approaches with greater frequency
than if learning opportunities are limited to when they present themselves
in the course of day-to-day interaction. According to Kevin Harvey, Colleen
Monahan, and Lars Ullberg:

Violence prevention is traditionally [a] very high stress conflict management situation. The Virtual World is removed from the face-to-face interactions, which has
several benefits. First, even though one can "experience" high-risk situations, there
is no real danger in a virtual environment. You can brandish a virtual weapon
without the real danger of getting shot. Second, it allows one to take advantage of
the abstraction of the Virtual World to unpack violent choices and violent behavior.
Third, it allows situations to be observed that are not observable in field conditions.
. . . Because there is no reliable way to hold the VI's [violence interrupters] account-
able for the intervention [because supervisors of violence interrupters cannot accompany them during an intervention without being a distraction and harming the
intervention], they cannot get feedback on their behavior or skills. In the Real
World, the only measure of success is the outcome—whether someone got shot or
not. In the Virtual World, not only can the scenarios be observed, but they can be
recorded and discussed at any time.[13]

According to a CeaseFire staff member, the use of Second Life for training has been minimal, however, due mainly to the lack of high-speed
computers and broadband Internet access in CeaseFire affiliate locations.
Colleen Moynihan, director of the CADE, sees this as a funding issue more
than a limitation of the technology. But CeaseFire has faced budgetary
challenges. The U.S. Department of Justice evaluation found that CeaseFire
sites had operating budgets of around $240,000 per site, but faced difficult
cuts in the summer of 2007.[14]

Another technological initiative planned by CeaseFire is to create events
data from SMS text messages, with these messages entered into a digital
map. Encouraging at-risk youth to send text messages will, it is hoped,
provide them with a degree of anonymity (so they are not regarded as
"snitches") while at the same time providing a communication channel
that can be timely in saving lives. CeaseFire is cooperating on this initiative
with Ushahidi (the crisis-mapping organization developed in Nairobi,

Kenya, as noted in chapter 2), PopTech, and FrontlineSMS:Medic in a project called PeaceTXT, with funding from the Rita Allen Foundation. According to the PopTech website: "The PeaceTXT project brings together technologists and social innovators to explore how mobile tools and mobile messaging might further accelerate CeaseFire's ability to engage communities, change social norms, improve its efficacy and find new paths to scale. . . . Experience and learning gained from this project is expected to benefit violence prevention efforts nationally and globally."[15]

Summary

The Chicago Project for Violence Prevention was founded by an public health physician in 1995, employing techniques for violence prevention similar to those used to eradicate diseases. In 2005, CeaseFire developed a program tailored to hospitals. When gang-related violence occurs, victims in emergency rooms are often accompanied by fellow gang members. Violence can break out if members of the opposing gang enter the hospital. Sometimes members of the opposing gang seek retribution for the injury or death of one or more of their members.

In 2007, CeaseFire received funding from the Robert Woods Johnson Foundation. The grant was designed to support the replication of CeaseFire in other U.S. cities. In addition, CeaseFire has affiliate projects in other countries.

At its core, CeaseFire is a community organizing enterprise that employs field monitors and interveners similar to the approach used by St. Xavier's. Technology supports that enterprise. Second Life is being used for training, and crowdsourcing and digital mapping approaches, using Ushahidi's platform, are being developed. At this point, it is not possible to measure the impact of using Ushahidi's technologies with CeaseFire's model of violence interruption.

External evaluators of CeaseFire noted the coherent theory of change that pervades the organization, based on the pulling of three levers: decisions, norms, and risks. The evaluation documented impressive reductions in killings and injuries resulting from CeaseFire's violence prevention initiatives in both neighborhoods and hospitals. As with St. Xavier's, Ceasefire's experience offers solid evidence that violence can be prevented through effective community organizing, a presence of trained people in "hot spots," and timely action when it becomes clear that tensions are escalating.

5   Counteracting Ethnoreligious Violence in Sri Lanka

This systemic and evidence-based approach to violence prevention at a micro-level has critical applications beyond Sri Lanka—particularly in areas of ethnic conflict or that are in the post conflict phase. Early warning for violence prevention can serve a critical role in hot-spots globally. . . . Imagine the potential impacts for overall national/provincial stability if such systems are deployed in the Northwest Frontier Province of Pakistan, the Nembe and Kalabari Kingdoms in Nigeria's Niger Delta, Pattani in Southern Thailand, or the post-conflict Malukus in Indonesia.

—David Nyheim, "Three Generations in Early Warning: Challenges and Future Directions"

We don't discuss the reasons why projects start strongly but then grind to a halt when the champion of that project moves to a new position, or the projects which deliver nice graphics but little operational value. As soon as the stakes are raised, and this is true for private and public organizations, there is a huge incentive not to discuss failure, or event to let it see the light of day. . . . We have an obligation to learn from those mistakes. . . . But to paraphrase [the comedian] George Carlin: scratch a skeptic, and you'll find a disappointed idealist underneath.

—Paul Currion, "Conclusion"

A good friend at The Asia Foundation called me one day to see if I would be willing to help him and others develop their conflict management and democratic governance programming. They had received a major grant to develop their capacity in this area from the Hewlett Foundation, and they needed someone to help them design their programmatic approach. I did periodic consulting work with The Asia Foundation from 2001 to 2006. This work took me to San Francisco (where The Asia Foundation is head-quartered), Bangkok, Thailand, and Kathmandu, Nepal. We developed "conflict diagnostics" with which to evaluate the impact of The Asia Foundation's programs.

In 2004, I was asked to conduct a midterm evaluation on a project in Sri Lanka, which The Asia Foundation was funding with support from the British High Commission. Over the next two years, I traveled to Sri Lanka to conduct the midterm evaluation, to monitor progress, and to provide technical support to the Foundation for Co-Existence (FCE). Over a two-year period, I learned more from those whom I evaluated than they learned from me—as is typical.

FCE faced financial challenges in 2009–2010 and discontinued its early warning and response program. Before that, however, I had the privilege of seeing this group of dedicated people grapple with the use of technology to prevent bloodshed in an extremely challenging conflict environment. Initially, they attempted to use an events data system that was developed for monitoring conflict and cooperation from a distance, in another country, not for monitoring localized violence. I saw how they created their own categories of events. I interacted with their program staff, statistician, computer programmers, event coders, field monitors, and volunteers.

In 2008 I traveled to Sri Lanka again to collect additional information as I was writing this book. My wife Sue came with me. We stayed at a hotel that looked out into the Indian Ocean. As she swam and I jogged on the beach in the mornings, we were reminded of the conflict plaguing the island by a machine gun nest atop a building across the street. Sue volunteered to write a report for a donor of a photojournalism initiative of FCE, called the Sahajeevana Centre for Coexistence. Pictures of violence on this beautiful island over the years were professionally displayed as a way of helping people process their emotions, overcome their hatred, and transform their anger. The theme of this exhibition summed up the general, tragic picture: "Paradise in Tears."[1]

Context

The minority Tamils, most of whom are Hindus, have long suffered from real and perceived discrimination on the "island of Ceylon," which is referred to as the "Golden Kingdom" by many devout Buddhists, and is now commonly known as Sri Lanka. Tamils reside predominantly in the North and East of Sri Lanka. In a protracted separatist revolt (often referred to as a "civil war") from 1983 to 2009, militant Tamils fought for independence from the majority Sinhalese, some of whom moved into the North and East as settlers, and most of whom are Buddhists. A ceasefire agreement between the government of Sri Lanka and the rebellious Liberation Tigers of Tamil Eelam (LTTE) was signed in February 2002, but violence between Tamil groups continued sporadically thereafter, substantially between Tamil Hindus (usually referred to as simply "Tamils") and Tamil Muslims (usually referred to as simply "Muslims") in the Eastern Province. The two communities suffered from significant tension due to past violence and concerns about equity in governance in a post-conflict political system. The Eastern Province is comprised of three districts: Ampara, Batticaloa, and Trincomalee.

In March 2004, the LTTE regional commander for Batticaloa-Ampara, Colonel Karuna Amman, broke away from the LTTE political-military command based in the North, claiming that Eastern Tamils were discriminated against by Northern Tamils in the administrative structure that the LTTE had created to govern areas under its control. Karuna's rebellion created a major split in the Tiger ranks. The rebellion was quickly crushed by the LTTE.

When Karuna announced the separation of the Eastern Province, people there celebrated in the streets, especially the youth. While Twitter feeds and other forms of conveying images of governmental brutality have turned citizens into on-the-spot reporters, as discussed in the Introduction, sadly, the technology was used for precision in torture and murder in the Eastern Province. Some LTTE loyalists videotaped the celebrations following Karuna's announcement. This footage was turned over to LTTE commanders. Soon thereafter, "disappearances" became common. It was not infrequent for those in the Eastern Province to find bodies of youth (especially young men) with their hands chained behind their backs, obviously tortured to death, lying on a street in the morning twilight. The Tigers wanted to make a point. They knew exactly who to torture to death to communicate that a separate Eastern Province was not an acceptable possibility to the LTTE. They had ample video footage with which to target unfortunate victims for months.

In view of the ongoing tensions in the Eastern Province, FCE was established in November 2002 with a mandate to support and strengthen coexistence between ethnoreligious groups. FCE initiated the Program on Human Security and Co-Existence. The British government, through The Asia Foundation, was the principal funder of this early warning and early response project. It began in April 2003. In following years, the government of Norway provided substantial funding as well. FCE was pulled into humanitarian relief in Sri Lanka following the tsunami on December 26, 2004.

Early Warning and Early Response System

FCE's Program on Human Security and Co-Existence consisted of an Information Center (referred to generally as the Information Technology Unit or "IT Unit") and a Program Unit. The former focused on developing an early warning system using events data. The consumers of reports generated from these data were local and national policy makers, diplomats, NGOs, INGOs, and the general public in Sri Lanka. swisspeace—an INGO based in Bern, Switzerland—provided training and technical support, sharing its Early Analysis of Tensions and Fact-finding system (known by the German acronym FAST) that used the IDEA events categorization and data analysis scheme.

The Program Unit developed capacity for early response. It created new or supported existing CBOs collectively called "Co-Existence Committees," or which went by other, more descriptive names such as "the committee of religious leaders." The Program Unit provided training for committee members. It set up offices in each district. A district coordinator supervised eight or so additional field officers. Staff members were selected carefully through a process of consultation with prominent community leaders (such as elders and religious leaders). Each field officer was expected to submit reports daily (either via fax or email) to FCE's Information Centre in Colombo about incidents of cooperation and conflict (a sample report format is provided in appendix A). Field officers and coordinators were balanced between Sinhalese (mainly Buddhists) and Tamils (mainly Hindus and Muslims, with some Christians as well). FCE field officers were required to submit at least one report of an event of the most extreme type of conflict or cooperation each day.

FCE conducted meetings, roundtable discussions, and district workshops to present their observations and analyses of the human security situation in the East. District coordinators were charged with the responsibility to respond to incipient conflict using mediation techniques in which they had been trained by FCE Colombo's staff. Usually, they responded to tension in collaboration with other organizations, such as the Sri Lanka Monitoring Mission (SLMM), an entity comprised of expatriates of Scandinavian countries charged with monitoring the ceasefire agreement. Sometimes they had assistance from FCE's Program Unit staff in Colombo.

The president of FCE, Kumar Rupesinghe, brought considerable early warning and early response expertise to the new organization. In addition to his academic credentials and publication record, he brought practical experience, having served as Secretary General of International Alert and as a founding board member of the Forum on Early Warning and Early Response (FEWER) in London. At FCE, in addition to developing the founding vision, he provided training for staff members in violence prevention.

Rupesinghe was kept informed of local developments. He contacted regional or national leaders, advocating intervention by them when he deemed their involvement essential to violence prevention, similar to what Prakash did in Ahmedabad.

In 2006, I directed a small team to write six case studies of FCE's work in the Eastern Province that was the basis of an impact assessment for the British High Commission. In addition to FCE staff members, I had the able assistance of Patricia Lawrence, an assistant professor at the University of Colorado at Boulder who had written her dissertation in Sri Lanka and was fluent in Tamil, and Timmo Gaasbeek, a PhD candidate at Wageningen University, the Netherlands, who also was fluent in Tamil after having lived in Sri Lanka for years. This chapter draws from these case studies.

General Observations

Here are some conclusions drawn from evaluations, monitoring reports, and case studies about the Human Security and Co-Existence program.

• Verification of data proved to be more time consuming than expected. FCE hired two former police officials who were charged with this responsibility.

• FCE used multiple sources for events data. In addition to reports of field officers and coordinators, FCE pulled data from seven newspapers, nine radio stations, one television station, and over a hundred websites in English, Sinhalese, and Tamil.

• There was unanimous agreement among current and potential users of the reports that they were useful in pulling together information that was otherwise scattered.

• FCE field officers consistently related that implementing the FAST system had numerous benefits. These included that they: (a) saw situations differently, both from the standpoint of looking for cooperation and conflict, as well as in noticing "lower grade" tensions; (b) were more diligent about assessing tensions on a regular basis; and (c) improved their analyses based on a greater amount and more precise categorization of information.

• On repeated occasions information gleaned from the "grass roots" was shared regularly with leaders at the mid- and top-levels, which showed that FCE had built a web of relationships vertically.

• The longer FCE used the FAST system, the more it became clear that it was not a good fit for their purposes. FAST was designed to assess the likelihood of conflict within a nation-state (a second-generation macrosystem), not at a community level. And there were limitations inherent in the FAST system that carried over into the one used by FCE. Namely, there was not an integrated pattern recognition capacity. The system, like that used by St. Xavier's, relied on inductive reasoning of staff and peace committee members to determine when violence was likely to happen and to develop a consensus that a warning be issued.

• FCE used only FAST until early 2006, when it started to run a parallel system of its own design, based on categories we developed in a workshop, listed in appendix B. At the end of 2007, FCE discontinued using FAST and relied exclusively on its own in-house system.

• As of March 2009, FCE had nearly 5,300 people at the grassroots level who served as Co-Existence Committee members. The Human Security and Co-Existence Program sent its reports to approximately 1,300 people, but only 90 received warnings via telephone calls, email, and text messages.[2]

• Doug Bond of Virtual Research Associates (VRA) assessed FCE's system in December 2008. He found that FCE did not employ automated pattern recognition (though staff members felt otherwise). His findings corroborated those of Lawrence, Gaasbeek, and myself that the Co-Existence Committees were extensive, with members willing to engage in early response (Bock, Lawrence, and Gaasbeek 2009).

Reports and Warnings

Over the years, FCE improved its Human Security Program reports substantially. Staff members decided to produce Daily Situation Reports, as well as monthly reports that summarized major themes and trends. Moreover, they substantially improved their collection of events data and reported the quantity of major events (such as number of child abductions and number of violent deaths) over time, depicting them with charts that were useful in assessing overall trends.

After writing case studies in the districts of Batticaloa and Ampara, Lawrence summarized what respondents said about FCE's Daily Situation Reports:

[They contain] more information than the other reports [i.e. of other NGOs], are not biased, and are based on [FCE's] . . . extensive networks on the ground, including Tamil, Muslim and Sinhalese who are effective and genuinely trying to get to the bottom of what is happening. On the other hand, . . . [others] were of the opinion that by the time Daily Situation Reports are put on the Internet, the situation has already shifted on the ground, so they are of use to people in Colombo or abroad, but not to people who are "walking on the sword" in their dangerous daily human rights work in the eastern districts. I was informed that some of the other NGOs in Batticaloa district had formed their own Internet circle to keep each other informed of local events. . . . They argued that it would be useful to organize a micro level system of rapid intra-district communications with organizations deescalating conflict, such as the Non-Violent Peace Force, the United Nations, the International Committee of the Red Cross and the Sri Lankan Monitoring Mission. (Bock, Lawrence, and Gaasbeek 2009)

We can see from these observations the importance of having staff members or peace committee members, or both, locally who communicate with staff members and committee members of similar organizations as a way of identifying incipient violence. When there is no automated pattern recognition capacity, linking people together locally can accomplish much the same thing. (One potential drawback to this approach, however, which is covered later in this book, is that there can be pathologies in group information processing and collective decision making. There can also be a lack of peripheral vision of what is happening outside their area.)

It was surprising how cell phones in general, and text messages in particular, became integrated into the utilization of data for early response. In the Batticaloa and Ampara districts, when there was an urgent situational change or event of violence, FCE team members sent very brief reports and received information back from headquarters using text messages.

There is a sense in which computerized events data combined with inductive reasoning can be beneficial when analyzed by those close to and away from the local conflict. Gaasbeek summarizes:

Data analysis at FCE Headquarters does support earlier response, but its exact impact is difficult to measure. This is because parallel [to] (and sometimes preceding) this analysis, information coming to district level staff and contact persons of FCE is informally analyzed and incorporated in these people's own mental framework of analysis. As the people in the districts live amidst the incidents on which they report, their analyses are much quicker and more grounded than any outsider's analysis could possibly be. However, being in the midst of a situation also means that there is a risk of becoming confused and losing track of larger patterns. (Bock, Lawrence, and Gaasbeek 2009)

FCE's Daily Situation Reports and data gathering were of value in informing organizations, institutions, and individuals in Colombo. They were of less value to those working in the field. What evolved over time, however, was that the discipline of collecting the information arguably enhanced FCE field staff members' awareness of what was happening. Through inductive reasoning, without the assistance of computerized pattern recognition, they made conclusions; communicated via text messages, cell phones and in person; and took action.

Results and Length of the Lull

How successful were FCE's interventions? As with any violence prevention effort, measuring impact is difficult. In August 2004, FCE provided me with a list of successful interventions in the Eastern Province. During field visits to the districts of Trincomalee and Batticaloa, I presented a description of those interventions to local people, government officials, and SLMM staff members as a means of verification. Four interventions were eliminated either due to my inability to verify them or because they were not viewed as distinct violence prevention activities. Only one of those incidents was singled out after an interviewee indicated that the event had happened but FCE had nothing to do with it. Otherwise, the consensus was that FCE had assisted in preventing violence, though it was one of numerous organizations that were instrumental in doing so.[3] I concluded that there were fourteen instances when FCE successfully prevented violence during its 2003–2004 start-up phase. This amounted to approximately one successful intervention per month.

In 2008, FCE staff provided me with another report, detailing their violence prevention efforts over a longer span. They listed 126 successful interventions, constituting an average of 2.1 per month, assuming the report was accurate.

Another way of measuring success is through statistical analysis. FCE coded, verified, and entered over 24,000 events from 2004 to 2008. Of that, data about conflict and cooperation between people of different ethnoreligious identities who were not a part of the larger separatist revolt on the island were a relatively small subset, comprised of approximately 3,700 events.[4] Here I report findings of tests done on these latter data only.

After purging data on the separatist revolt from our database, my research assistant, Robert Perera, and I aggregated events into categories similar to those used by Meier, Bond, and Bond (2007), which we called "super events." Cooperative categories included INITIATIVES, MITIGATION, EXCHANGES, and AID. Categories of conflict included WARNINGS, AGGRAVATORS, PROVOCATION, PRESSURE, CONSENSUS BUILDING FOR VIOLENCE, and VIOLENCE. A complete listing of the super events, main events, and subevents for these categories appears in appendix C.

We used FCE's data on PROVOCATION as a surrogate for a precipitating event as defined by Horowitz (2001, 269). Because of the judgment involved in labeling an event as such, I asked a staff member in Colombo to double-check accuracy of those data. So, not only was each event coded and verified initially, but also each entry for provocation was checked again to make sure it was a "symbolically potent" event.[5]

In measuring the efficacy of violence mitigation with FCE's data, we used two types of data for VIOLENCE. The main event category of HUMAN CONDITIONS, with the subevent HUMAN DEATH, is an indicator of whether there was a loss of life during the event, making it *nominal* data, a measure of *incidence*—that is, whether there was violence or not. In contrast, *interval* data—that which measures amounts with precision between those amounts, as in the number of people—included counts of HUMAN DEATHS. These measured the *intensity* of violence.[6]

With nominal data, we analyzed if there was any difference between a sequence of provocation—mitigation—violence as compared to a sequence of provocation—violence. We identified thirty-six cases in which mitigation followed provocations. We found that violence followed mitigation twenty-six times, whereas no violence followed mitigation ten times. But this was not a statistically significant result.

We then tested the efficacy of mitigation relative to the *intensity* of violence. The data contain events for 2,391 violent deaths during the period of our analysis. We derived a statistically significant finding that the number of deaths for the sequence provocation—mitigation—violence had a mean (an average) of 1.33 human deaths, whereas when the sequence was provocation—violence the mean was 2.13. The difference between 1.33 deaths and 2.13 deaths is statistically significant.[7] This indicates that violence mitigation undertaken by Co-Existence Committees, FCE staff members, and other actors (such as police and religious leaders) reduced the lethality of deadly violence between groups of different ethnoreligious identities by roughly 23 percent.

I was also curious about the amount of time that elapses between a precipitating event and violence—what Horowitz (2001) calls the "lull." To test this, we sorted the data by day and measured how many days were between the super events PROVOCATION and VIOLENCE (we did not have data on hours). There were 2,476 cases involving PROVOCATION and VIOLENCE in all. After removing outliers, figure 5.1 shows the histogram of the observed length of the lull.[8] We found that the average length, the mean, of the lull for violence between groups of different religious identities in this analysis is 2.88 days.[9] The median, the most frequently

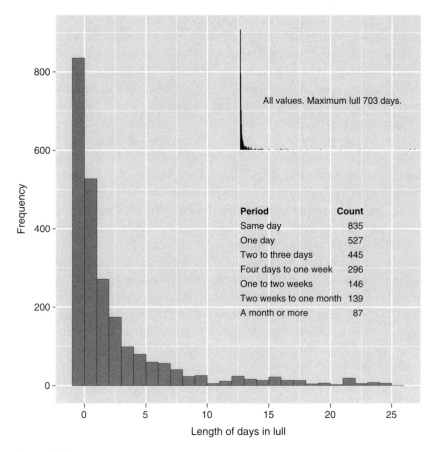

All values. Maximum lull 703 days.

Period	Count
Same day	835
One day	527
Two to three days	445
Four days to one week	296
One to two weeks	146
Two weeks to one month	139
A month or more	87

Figure 5.1
Length of the lull

occurring observation, is 1 day. Specifically, we found that violence erupted 35 percent of the time during the first day, 60 percent during the first and second days, 68 percent during the first, second, and third days, and 76 percent during the first, second, third, and fourth days. These findings support Horowitz's observation that the lull is typically "a matter of hours or days" and generally not weeks. It indicates clearly that a typical lull is not a matter of months.

Research Limitations

I want to be clear about the limitations of our data for Sri Lanka and stress that the findings reported here are preliminary. There are five limitations worthy of mention. First, the data are from a conflict between people of different ethnoreligious identities during a larger separatist revolt that had an ethnoreligious identity dimension, thereby limiting the relevance of the findings. The dynamics of violence along ethnoreligious lines not influenced by a larger separatist revolt could be very different.

Second, the events were not linked substantively. We used geographical proximity as a proxy for linkage of events. We made an assumption that a precipitating event in an area the size of a district is likely to become known within that district and, hence, that geographical space (i.e., district) could reasonably serve as a proxy for linkage in event sequencing. We did not have data on the relationship of events to one another. In my conversations with FCE staff members, however, they assured me that district data are the best geographical area for event linkage. They did say that sometimes events in one district "bleed" into other districts but argued that such instances are rare. Because of this, we eliminated events for which the district measure was imperfect. Namely, we purged four events that were marked "all provinces," another 254 events for which the province and district were labeled "none," and, finally, another 164 events that had the province listed but not the district.

Third, another limitation is that—again, because we did not have events linked substantively—we were unable to separate out those events related to two-wave and recurring violence. Since Horowitz (2001) considers the lull for two-wave and recurring violence to be longer than nonrecurring violence, this lack of separation probably skewed the result somewhat in the direction of a longer lull overall.

Fourth, there is no established method to determine the length of a lull when multiple provocative events occur sequentially. We faced the question of what to do when there were multiple provocations without violence in between. Does that mean the lull began with the first provocation or the last? We decided to measure the length of time between the most recent provocation and violence. We assumed this was preferable because not all precipitating events result in violence.

And, fifth, the sample is not randomized. A causal inference, of course, can only be made when there has been randomization of the sample. In measuring the efficacy of mitigation, intervening or not intervening would have to be done randomly. This, of course, would violate social science ethics.

So, in brief, we need to be content that this analysis gives evidence that the lull between a precipitating event and violence along ethnoreligious lines within a larger separatist revolt is usually of a relatively short duration (as in hours or days). Moreover, efforts at mitigation in such situations are associated with less intensity of violence as compared to instances when there are no efforts to prevent violence.

Summary

FCE was established in November 2002 with a mandate to support and strengthen coexistence between ethnoreligious groups. FCE's Program on Human Security and Co-Existence consisted of an Information Center (referred to generally as the Information Technology Unit or "IT Unit") and a Program Unit. swisspeace, an INGO based in Bern, Switzerland, provided training and technical support, sharing its FAST system that used the IDEA events categorization and data analysis scheme.

The Program Unit developed capacity for early response. It created new or supported existing CBOs collectively called "Co-Existence Committees." The Program Unit provided training in mediation for committee members. It set up offices in each district, a subunit of a province. District coordinators were charged with the responsibility to respond to incipient conflict using mediation techniques in which they had been trained by FCE Colombo's staff. Usually, they responded to tension in collaboration with other organizations, such as the Sri Lanka Monitoring Mission (SLMM).

FCE's experiences shed light on an early warning system that assembles events data while using inductive reasoning to issue a warning to invoke an early response. FCE found that verification of data was more time consuming than expected. The organization's reports were considered useful by consumers within government, diplomatic, and NGO circles in that they pulled together information that was otherwise scattered.

FCE field officers did not find the reports to be helpful for early warning purposes, but they did relate how the training they had received, and the categories about which they had to report, helped them refine their analytical capacities, and made them more diligent about assessing tensions. Through inductive reasoning, without the assistance of computerized pattern recognition, they made conclusions; communicated via text messages, cell phones, and in person; and took action.

On repeated occasions information gleaned from the "grass roots" was shared regularly with leaders at the mid- and top-levels, which showed that FCE had built a web of relationships vertically.

As of March 2009, FCE had nearly 5,300 people at the grassroots who served as Co-Existence Committee members. The Human Security Program sent its reports to approximately 1,300 people, but only 90 received warnings via telephone calls, email, and text messages.

As for results, according to FCE's own analyses, it intervened successfully to prevent violence approximately once a month during the start-up phase in 2003–2004, and 2.1 per month from 2004 to 2008.

In analyzing FCE's events data, we found a statistically significant result that efforts to prevent violence were successful part of the time. We also found that the average length of the lull between provocation and violence was 2.88 days. The median was one day. This supports support Horowitz's observation that the lull is typically "a matter of hours or days" and generally not weeks (2001). It indicates clearly that a typical lull is not a matter of months.

6 Crowdsourcing during Post-election Violence in Kenya

Ushahidi . . . represents a new frontier of innovation. Silicon Valley has been the reigning paradigm of innovation, with its universities, financiers, mentors, immigrants and robust patents. Ushahidi comes from another world, in which entrepreneurship is born of hardship and innovators focus on doing more with less, rather than on selling you new and improved stuff.

Because Ushahidi originated in crisis, no one tried to patent and monopolize it. Because Kenya is poor, with computers out of reach for many, Ushahidi made its system work on cell phones. Because Ushahidi had no venture-capital backing, it used open-source software and was thus free to let others remix its tool for new projects.

—Anand Giridharadas, "Africa's Gift to Silicon Valley: How to Track a Crisis"

In a crisis there is a relentless and unforgiving trend towards an ever greater information transparency. In the most remote and hostile locations of the globe, hundreds of millions of electronic eyes and ears are creating a capacity for scrutiny and new demands for accountability. . . . The moment any crisis incident takes place there is an imperative to fill the resulting information space within not hours but minutes, and if possible to dominate it.

—Nik Gowing, *Skyful of Lies and Black Swans*

The availability of GIS (geographic information system) data has increased the potential for greater precision in early warning and early response at a local level. Indeed, the uses of GIS and events data go far beyond violence prevention. GIS data, combined with information on rainfall and the like, are being used to predict famine (Logan and Moseley 2001). GIS and health data are being used by epidemiologists for infectious disease control. While country-wide data might indicate a very low incidence of, say, tuberculosis per capita, linking infectious disease data over time with GIS data can identify concentrated pockets of a disease where it is spreading exponentially. This is similar to violence prevention. While a society might be managing tensions between groups in peaceful ways, there are often pockets of trouble, where violence can ignite and spread rapidly.

Georeferencing events data creates the potential to identify violence or potential violence at a specific location. Of course, the more narrow the geographic scope of an analysis, the more difficult it is to use pattern recognition, especially if relying on field staff for data. In contrast, crowdsourcing approaches, if many people participate, will typically yield more data and, hence, make it more plausible to have enough information for pattern recognition within a small geographic area. This means that using crowdsourcing can make mathematical pattern recognition more feasible for providing actionable information. The location of the problem will be more specific and, therefore, more useful for early response.

As we saw with FCE, events data can be developed by field officers. But that requires people working day after day to observe, report, validate, and enter the data into a computer system. This is costly in both human and financial terms.

With crowdsourcing, data can be gathered from text messages and reports sent online. Combining them with blog postings and other Internet-based text and analyzing them can constitute what Cass Sunstein calls "prediction markets." He summarizes his findings of this type of "collective intelligence" by stating:

> At their best, the new methods have two remarkable virtues. First, they show us fresh ways to obtain access to the information held by many minds. Second, they show us how we might dramatically improve the old method of deliberation by increasing the likelihood that groups can learn what their members know. Access to many minds contains risks, because many people can and do blunder. But . . . dispersed information, if elicited, is far more likely to lead to better understanding—and ultimately to more sensible decisions in both markets and politics. (2006, viii)

Prediction markets linked with dynamic mapping has the potential to dramatically improve early warning and early response capacity for preventing local violence. According to Meier, this involves "neogeography" whereby people create their own maps. This approach was facilitated substantially when GoogleEarth™ was launched. The Peace Research Institute in Oslo (PRIO) georeferenced historical conflict information and, with the help of some universities in Europe, produced GROW-net in 2006. A year later, the Swiss Federal Institute of Technology, one of the collaborators in developing GROW-net, used GoogleEarth™ for crisis mapping.[1] The project, called WarViews, offered two kinds of data, which the Switzerland-based researchers euphemistically called "flavors." Their aim is to create an easy-to-use front-end for the exploration of GIS data on conflict. It takes advantage of the recent proliferation of [I]nternet-based geographic software and makes geographic data on conflict available for these tools. . . . WarViews comes in two "flavors": The static version runs in a web browser and allows the user to switch between different datasets. The dynamic version is based on GoogleEarth™ and can animate geographic data such that the development over time can be monitored.[2]

The second "flavor," according to Weidmann and Kuse, "allows for the examination and discovery of spatio-temporal patterns of conflict" (2009, 37).

A leading technology provider for this kind of approach to early warning at a local level is Ushahidi, which takes its name from a Swahili word meaning "testimony." Ushahidi was created initially by journalists monitoring violence in Kenya following the December 2008 elections. Sokari Ekine, a human rights and digital activist, describes the process:

Within 24 hours of the outbreak of the 2008/2009 post election violence in Kenya, Kenyan blogs were posting hour-by-hour reports. On 31 December, there was a complete shutdown of the mainstream media. . . . Within days, the online community and blog aggregator, Mashada, had set up an SMS and voice hotline calling for people to send in local news and opinions on what was happening. This was followed by Ory Okolloh (Kenyan pundit) who suggested using Google Earth to create a *mashup* [a website that aggregates data in useful ways] of where the violence was taking place and called upon "any techies" out there willing to help create a map of it. This was 3 January and by 9 January a group of Kenyan bloggers had put together a mashup and created Ushahidi, a site for people to send SMS or email reports of acts of violence directly. (2010, xix; emphasis added)

The founders of Ushahidi formed a nonprofit technology company that supports numerous emerging volunteer and technical communities (V&TCs)—part social movement, part NGO—which creates a platform to enable volunteers and staff members to contribute to information flow related to a specific problem, such as a disaster or a potential riot. The journalists felt a need to have input from the community on significant events during the crisis. Since then, Ushahidi adopted a goal "to create a platform that any person or organization can use to set up their own way to collect and visualize information."[3] The design is flexible. Users can download the "core platform" (free) and adapt it as needed. Staff members at Ushahidi provide guidance, using wikis and Skype chats where users can help themselves as they develop an application. Such platforms have been created in these situations, among others:

- the Gaza-Israeli military confrontation in early 2009;
- violence in the Democratic Republic of Congo (DRC);
- attacks perpetrated against non-South Africans in South Africa;
- elections in India in 2009;
- elections in Afghanistan in 2009;
- disaster response to the massive earthquake in Haiti in 2010;
- conflict mapping in Libya following the February 2011 rebellion; and
- response to emergencies and radiation challenges in Japan following the earthquake in April 2011.

Ushahidi's software is "open source," allowing volunteers and staff members to improve the program over time, collaborating remotely over the Internet. It has programming volunteers and staff members from throughout the world in such disparate places as Ghana, Kenya, Malawi, the Netherlands, South Africa, and the United States. The organizations that use Ushahidi's software use crowdsourcing, by which information is collected directly from as many people as possible near or in the area of concern. This is an approach used by Wikipedia, for instance, in harnessing the collective writing of many contributors into an online encyclopedia. In the case of organizations using Ushahidi's technology, however, individuals directly submit information on events by filling in a form on the Internet or by sending a text message. Data are also pulled from social media, including Twitter, Facebook, and blogs (Heinzelman and Waters 2010, 2, 7). Extracting information from social media and other Internet-based sources is sometimes referred to as *data harvesting*. This can be done manually. As people see information about conflict and cooperation, they can then place the event on a digital map (if information on location is available).

Ushahidi's software depicts events acquired over time on a web-based map, once those handling the information flow determine the location of an event. It provides a platform with which to send alerts to information contributors and others. Recipients subscribe to receive notices of event types in specific locations. The site for the Democratic Republic of Congo, for instance, asks in what city the recipient lives and allows him or her to receive alerts via text message, email, or Really Simple Syndication (RSS) feeds. This, as aptly stated by Meier, "closes the crowdsourcing to a crowd feeding feedback loop."[4]

Of course, *crowd feeding*, sending warnings to those who have contributed events data, could be used to incite violence as well. Information that is posted, however, has to be moderated beforehand, which reduces this risk.[5]

How do people find out about Ushahidi and begin to use it to report events? In attempting to explain, Erik Hersman, director of operations for Ushahidi, writes, "I don't have . . . a complete answer other than if it . . . [is] up and live, it . . . [will] gain traction over time, just as any effective web/mobile service does."[6] Some V&TC members have found that disseminating information about how to participate is best done through Facebook and Twitter (Salazar and Soto 2011, 61).

In India, Ushahidi software was downloaded by election observers concerned about potential corruption and election-related violence. The group, Vote Report India, asked observers to report within categories of events, using a "shorthand" label so a computer would recognize it, with a brief explanation. These labels and definitions are taken verbatim from their website:

• Election Commission Interventions (ECIN): Election Commission interventions in response to violations of the Model Code of Conduct [which was promulgated by the Election Commission of India in 2007, providing guidance about appropriate campaigning].[7]

• Forged Vote (FORG): Voter being unable to cast her/his vote because a third-party has voted by fraud or impersonation.

• Inflammatory Speech (HATE): Spoken or written speech aimed at inciting communal differences along the lines of religion, caste, or creed.

• Other Irregularities (OTHR): All other violations of the Model Code of Conduct not included here.

• Violence (VIOL): Violence related to the elections: voter intimidation, interparty clashes, mob violence, [voting] booth capture.

• Voter Bribing (BRIB): Voter bribing through the distribution of cash, alcohol, food, gifts, etc.

• Voter Name Missing (MISS): Voter's name missing in spite of her/him having completed due procedures.

• Voting Machine Problems (EVOM): Electronic voting machine irregularities: malfunction, power outage, [stuck button], etc.

• What Went Well (WELL): The electoral process worked as it should.[8]

There were four ways in which Vote Report India received events data. The first was via a text message that started with VOTERREPORT. A second was via an email message to report@votereport.in. The third: an election observer could send a Twitter message beginning with the number sign (otherwise called a "hash-tag") #votereport. And, finally, an event could be reported by filling out a form on the Vote Report India website.

Text messages received would have to contain an event code. Here is an example that Vote Report India provided on its website:

votereport #Pune #VIOL Violence in Pune's Boat Club area. 6 injured. Situation under control. Source: NDTV.

Here is another example:

votereport #400019 #BRIB Liquor and clothes distributed in Dharavi slum on election even by X candidate. Eyewitness account.[9]

A Ushahidi development team is creating a capacity for incoming reports to be categorized automatically using *Bayesian modeling*. This is an approach whereby clusters of words are used to establish categories that are honed by developing and refining hypotheses of the likelihood of association. The software, once developed, will look for "clues" in each report to determine the most likely event category. So, for instance, if a report were submitted in which the words include "mob," "Molotov cocktail," "Hutu," and "Tutsi," the category would be categorized as *riot*; whereas if the report included "mob," "Hutu," and "Tutsi," with "mob" but without "Molotov cocktail," the event would be categorized as *protest*. Over time, clusters of "clues" are assembled, making categorization more accurate. Imagine how much easier it would be for people to submit reports to Ushahidi if they did not have to classify them. Rather than having to list the category, a cell phone user could simply scroll down for the number of where to send the text message and then send a regular message.[10]

Concerns about Data Validity

One of the concerns about crowdsourcing by organizations that use Ushahidi's technology—voiced by those attending a conference in Karachi, Pakistan, that I attended in March 2009—is that text messages and Internet postings are sometimes manipulative. They can be designed by troublemakers and spoilers to create panic or concern. So aggregating manipulative communications will give a false reading, and to impose a verification process requires time, thereby impeding information flow, not to mention the need to expend both human and financial resources. Of the 3,500 messages reported by Ushahidi during the response to the Haitian earthquake in January 2010, Heinzelman and Waters report that "only 202 were tagged as 'verified,' mostly from early Web submissions that had been based on media reports" (2010, 11). In other words, only 5.77 percent were verified, though we must keep in mind that Ushahidi's purpose in Haiti was for crisis management, not early warning. Hence, it was not especially political and the likelihood of manipulative submissions was minimal.

When Ushahidi's software is used by organizations engaged in *conflict transformation* (a process that does not necessarily bring resolution of conflict but that develops approaches for dealing with it constructively, with an assumption that conflict is inevitable and not always undesirable but that violent conflict is undesirable), such as the youth movement in Egypt in 2011, verification becomes more important. In this case, 91 percent of the 2,700 reports submitted were verified.[11]

It is important to keep in mind that numerous web-based approaches to information and knowledge sharing typically have verification challenges. Take, for instance, Wikipedia. People can and do post erroneous content on a subject. Such postings misguide users and spread false information. Hence, a quality-control mechanism or process of sorts is required. Ideally, the online "community" itself will undertake this, which can involve correcting or deleting false information or responding to an accusation. The latter sometimes happens when posting accurate information creates pressure for change. For example, if a company is accused on a site like YouTube of selling a faulty product (as was a bicycle lock corporation when a boy posted a video showing how he could open one of that company's lock models with a ballpoint pen), then the company has the opportunity to either correct the posting as lacking credibility, on the one hand, or it can pull the product from the market in a recall, on the other. In this case—it actually happened—the bike lock company recalled the boy's model (Zarrella 2009, 4).[12]

Since time is often of the essence in conflict early response, checking data accuracy quickly is essential. For example, a posting during a cease-fire that a renegade militant group plans to attack a village could either be true or a fabrication designed to instill fear among villagers. A false alarm could cause panic. But not responding right away, if the data are true, could mean the difference between life and death for hundreds, if not thousands, of people.

So how is Ushahidi verifying data quickly? Ushahidi offers numerous approaches, either being developed or already in use. One approach is to appeal to morality. According to Meier, this involves asking those submitting data to "check-off that the information they submit is true." This attempt at facilitating *crowdsourcing honesty* was inspired by behavioral economics. Specifically, Dan Ariely's research (2008, 212–213) has shown that MIT students were far more honest in their answers to questions when they were "reminded of morality."[13]

Another approach is mathematical, involving a sort of "peer review" process along with measures of reputation, redundancy, and proximity. The peer review process is a remarkably simple but useful way to check the validity of crowdsourced information. And this can be done in near-real time. One way of doing so, as explained by Meier, is to add a flag on data points on the digital map where the information is displayed. If a data point has a yellow flag, it means it is unverified data. The crowd can then take a look at it. If they agree that the event happened, they can replace the yellow with a green flag. If they disagree, they can make the flag red. With multiple participants in a crowd, these verification data can be aggregated to yield an overall peer-reviewed result. That is, if most of the "peers" award an event a green flag, it is considered verified or, conversely, if the majority of peers give a report a red flag, the data are considered unreliable. Ushahidi developers call this approach to verification *crowdsourcing the filter.*[14]

The mathematical approach for "reputation" is based on prior experiences with the person posting data by determining whether past information was corroborated or contradicted.[15] For instance, if a person's report receives an aggregate green flag through peer review in the past, then data submitted by that individual in the future is considered likely to be reliable. A score is ranked higher if the report comes from the location of the event, or near to it. In addition, it is ranked even higher if there are duplicate reports of the same event (unless it is exactly the same, which suggests that someone is "working the system").

Ushahidi is attempting to use automation to check redundancy. That is, Ushahidi software developers are using *predictive tagging* to dissect reports into data on the "who, what, when, and where" of an event. Then data are compared to determine which reports are about the same event. These are "tagged" as being virtually the same. This has the dual advantage of preventing double counting of an event while also adding credibility to the reports that corroborate.[16]

These four mathematical approaches—peer review, reliability, proximity, and redundancy—can be combined to reach an overall score, which the Ushahidi team calls a *reality score.*[17] Ushahidi staff members refer to these approaches collectively as a *digital straw* (think of a soda straw)—software that pulls usable data from an *information fire hose.* Such a fire hose has increasing information in both volume and speed. The digital straw, therefore, makes it possible, ideally, to have enough confidence in the veracity of some of the data to subject it to analyses with enough confidence to "distill a *signal* from the *noise.*"[18] Verified data show up on the map while data lacking credibility "fade away."

A more "low-tech" approach to checking validity of data is to send a message back or to call those who submit reports that seem dubious, asking for more information. Or organizations using the Ushahidi platform can check validity by communicating with their networks of contacts. In some versions of Ushahidi, reports are "approved" and "disapproved." In the process of making that determination, a message is sometimes sent back to the source using Clickatell, a text-message gateway that allows users in some locations to economically send messages elsewhere, including other countries, inexpensively.[19]

Another "low-tech" way of enhancing validity of data is to develop a *trust network*. These can be volunteers or paid staff members, or both, who constitute a trusted group of reporters. The information they provide can be combined with that coming from people outside the network. The crowd, therefore, is still involved, but those in the network can do rapid, on the spot verification checks. This approach is called *bounded crowdsourcing*. It involves combining data from trusted sources with data received from the general population using a hybrid approach (Heinzelman and Waters 2010, 12). The advantage of this method is that trusted sources can help with rapid, near-real-time verification or provision of data when time is of the essence, while still allowing for input from a wide group of people. Members of the trust network can also expedite the provision of information about the location of an event, and can intervene to prevent violence as needed.[20]

Nefarious Actors

As I explained with FCE, there is trepidation about allowing open access to early warnings. Ushahidi staff members have a similar concern. They do not want their digital maps to be used by those promoting violence. One solution that has been proposed is to create secure web pages.[21] This, of course, works at cross purposes with an approach that uses crowdsourcing. Hence, there has been a fascinating exchange of ideas about what to do regarding use of the Ushahidi platform by those interested in promoting rather than preventing violence. This is not unlike FCE's careful selection of to whom to send reports and, even more constrained, to whom to send early warnings. This concern was described in a blog post by Erik Hersman, Ushahidi's Director of Operations and Strategy, regarding the terrorist attacks in Mumbai, India, in 2008:

As much as mainstream media and experts are up in arms over the way that the terrorists in Mumbai could use information coming in from these new digital channels to monitor their own situation, we have to remember this isn't new. Groups like this have been able to do this with mainstream TV and radio for years. What's disturbing is that not even the government can stop it now.

The problem is that it's no longer one-to-many mass broadcast, it's now mass-broadcast to mass-broadcast. How do you stop six million SMS messages without crippling your own infrastructure and ability to get work done?

I think one answer might be found in figuring out a way to harness information from an even greater number of people. The more data that is collected, the less chance that bad data can have an adverse effect. For instance, if two reports come in that widely differ from the reports by ten other people, then we can assume that they are false. That at least helps us solve for a greater probability of good info being available and can help with the adverse use of it by the "bad guys."

What it doesn't do is solve for the problem of the "bad guys" having more information available at their fingertips. Nothing will solve that now. What it does do is mean those opposing them will have equal access to the same information, and possibly even more than is currently available on the "bad guys'" movements and operations by tapping into the greater public.[22]

One response to this post on Ushahidi's blog was: "The 'bad guys' in Mumbai have been reported to have used their [B]lackBerry's and [G]oogle maps. I believe this is an unintended consequence of the democratization of information. This [sic] examples should not stop the rest of us from using information for as much good as we can. I don't think worrying about the bad guys is worth our time."[23]

There are at least four ways to view or respond to the problem of creating crowdsourcing platforms that are used for nefarious purposes such as violence cultivation. One is to assume those promoting violence will have useful digital mapping tools at their disposal anyway and to presume that Ushahidi is providing a service that can at least provide those trying to prevent violence with a useful system. A second is to develop ways in which a data-filtering process discards data entered by those who are using the site for nefarious purposes (though that is a challenging type of digital straw to develop). Third, secure communications can be used in "nonpermissive environments" as when civil liberties are constrained (Meier 2011, 14). And, fourth, is to pay attention to the information that the extremists are submitting in case it portends their plan of action, which would enhance early warning and early response capacity.

It is important to differentiate between those individuals or groups who are likely to use social media and crowdsourcing platforms for gaining or holding dominance and those who will otherwise spread information deceptively, as was done in Iran in 2009. Renegade elements, such as militant branches of political parties with large followings, are likely to be more successful in creating panic and hysteria than, say, a governmental intelligence agency, a guerrilla group, or an organized crime organization. As Meier points out, the latter types of organizations have "highly centralized bureaucratic structures that may prevent them from being as innovative in their application of ICTs compared to more distributed networks (2011, 13).

We have very little understanding of how successful different actors will be with different approaches. A distinction should be made, for instance, between cultivating panic through rumors to create a sense of insecurity in a high-stakes election cycle in a parliamentary democracy, on the one hand, compared to disseminating propaganda to discredit opponents as a way of keeping a repressive regime in power, on the other. There is a sense in which rumors can be propagated "organically," whereas propaganda is often used in relation to a relatively well thought-out strategy.

Limited Internet Access

A brief case study on the use of Ushahidi for mapping "Peace Heroes" in Kenya was written by Melissa Tully on behalf of a social change–oriented design organization based in Amsterdam called Butterfly Works. After downloading a copy of Ushahidi's software, she created a website in collaboration with the Media Focus on Africa Foundation (MFOA).[24] While this was not intended to be a conflict early warning system, it became one. Tully interviewed a Ushahidi user, Marten Schoonman, Projects Coordinator at MFOA. He said that solicitation of positive examples of peacebuilding inadvertently provided an "electronic venue" for reporting about potential violence. Those, too, were mapped on the Peace Heroes website.

Schoonman points out that although cell phones are ubiquitous, computers with high-speed Internet access are of limited availability in many parts of Kenya. Therefore, the network of people who send information are not necessarily benefitting from the graphic portrayal of aggregated events. He also found that people need to be drawn in to the project. Due in part to the limited availability of computers online, there was no "viral effect" in getting more people involved.

Given the limited Internet access in Kenya, Schoonman found that passing out flyers at peace-related events was the most effective way to attract more participants.[25] The flyers advocate that people get involved in reporting incidents of cooperation and conflict by appealing to civic consciousness.

There are numerous ways in which people communicate even without Internet access. It is estimated that roughly 5 percent of Kenya's population had access to the Internet at the time of the December 27, 2007, presidential election that resulted in a violent aftermath. But among the 5 percent were those who posted information on blogs. Radio stations verified the information by checking other sources. If verified and important, program hosts read the blog posts over the radio; approximately 90 percent of Kenyans have access to a radio (Meier and Leaning 2009, 10).

Mapping Rumors

One of the applications of Ushahidi for violence prevention is to map rumors. As we saw in the case of St. Xavier's in chapter 3, rumors are helpful in identifying when consensus toward violence is building. Tracking them, therefore, can be a useful early warning approach.

In Lofa County, Liberia, for instance, when a young Christian woman was murdered, rumors spread that the local imam was to blame. A group of student protesters threw rocks at the imam who was inside a mosque. The imam's son then called relatives and told them his father was killed and the mosque was burned (neither of which were true). Muslims then burned down churches.

How might Ushahidi have been useful in this situation? One observer of the mob violence, a member of the Crisis Mapping Network, asks, "What if these rumors were tracked, documented in near-real time? Would we then have a new list of indicators for increased risk of violence and could these indicators be the early warnings that crisis mappers have been waiting for? The revolution that has yet to take place is one where rumors are considered valuable information, and not simply invalid weapons of the uneducated or irrational." She adds that "there are plenty of dangers in mapping rumors, one being that anything that appears on the map may be considered verified despite its categorization as otherwise. And if viewers consider the rumors validated, the mere presence of rumors on the map could indeed lead to acts of retribution stemming from the Ushahidi instance rather than the rumor on the ground. These are plenty of good reasons to be cautious."[26]

Posting a digital map of rumors on the Internet would be like FCE sharing its events database and early warnings publicly. FCE did neither. Both were restricted. While crowdsourcing increases the volume of data, crowd feeding—both in posting information on a digital map on the Internet and in sending warnings—can bring unintended consequences of exacerbating rather than reducing tensions. This does not mean, of course, that mapping rumors is ill advised. Doing so arguably is one of the most effective ways of identifying when and where violence is likely to happen. One must recognize, however, that there is an inherent transparency conundrum in approaches that seek and acquire information from "the crowd" while disseminating insights drawn from that information to "a few."

Communicating Does Not Necessarily Mean Intervening

There is every reason to believe that generating data for early warning will be increasingly robust in the future. Less clear, however, is the extent to which this will result in early response. Will more ubiquitous early warning systems result in the more effective utilization of information they provide to save lives?

One troubling observation, offered by Molly Beutz Land, is that there tends to be an "inverse relationship between meaningful *participation* and *mobilization*" (2009, 223; emphasis added). Through interviews of some of the pioneers of "netizen" democracy (to use a term coined by Michael Hauben [Hauben and Hauben 1997]), Land found that even when people contribute to an information pool about a social problem, not many get involved in doing anything about it. She found that "only a small percentage mobilized develop a sustained commitment to the cause."[27] Avaaz, a web-based citizen activism platform focused mainly on human rights and environmental causes, has approximately 3.5 million members. But, according to its executive director, Ricken Patel, only somewhere between 10 and 15 percent of those members donate or attend rallies. An even smaller percent of members are involved in making decisions about the direction of the "movement" (Land 2009, 223). We need to be careful, of course, not to draw conclusions about crowdsourcing for reporting human rights violations compared to crowdsourcing for reporting about existing or likely violence. The former is typically perpetrated by governmental entities, rebel groups, or terrorist organizations while the latter is often performed by collections of people at a local level who are not necessarily linked to a larger organization. The perception of organizational depth is one factor impacting the propensity to intervene to prevent a human rights violation, on the one hand, or to prevent a violent incident, on the other.

Two potential ways to increase mobilization for violence prevention resulting from participation in a crowdsourcing process are to include more people in decision making and to mobilize through groups. At the same time, giving large numbers of people a say in decisions about what to do with the information is challenging. Even organizations like Amnesty International and Wikipedia, which have millions of members or participants who write advocacy letters (Amnesty International) or contribute to an online encyclopedia article (Wikipedia), have a very small number of people deciding what issue to campaign on or whether a given entry is acceptable to post on the website (Land 2009).

The other potential way to overcome the gap between mobilization and action is to facilitate participation through the formation of groups that hold collective identities. This, according to Michael D. Ayers, requires that the groups develop an ability to create a "shared definition system" through "a series of self reevaluations of shared experiences, shared opportunities, and shared interests" (2003, 145, 151–52, as quoted by Land 2009). This observation is reminiscent of Paulo Freire's guidance on how to raise critical consciousness. Namely, doing so requires awareness, reflection, and action (1997, chapter 3).

An example of creating groups with members who are capable and willing to engage in both early warning and early response are the peace committees facilitated by St. Xavier's and FCE. Supporting these CBOs required in-person conversations, meetings, training activities, and engagement to take action at times of high tension. Compare that with receiving a brochure at a rally or getting a message urging participation in citizen action to prevent violence or human rights abuses. Participating as a reporter of events is one thing. Working collaboratively with a group to intervene in a volatile, often dangerous, situation is quite another.

The gap between providing information, on the one hand, and acting effectively on the basis of that information to mitigate violence, on the other, mirrors the fundamental early warning and early response challenge. Namely, there tends to be a gap between information and action. So as ways of getting more and more people to contribute information are being developed, as Ushahidi has done successfully, it is important to implement deliberate measures to foster participation in early response.[28] This could take the form of recruiting and training peace committee members who have a sense of belonging to their respective group, whose members are provided timely early warning information and are prepared and willing to undertake early action. Doing so suggests that those using *bounded crowdsourcing* should also engage in *bounded crowd feeding* or *restricted feeding* (in the case of warnings). While the former provides a solution to the need to verify data in real time, the latter offers a core group of trained people who can intervene to prevent violence and are capable of being prudent with sharing information that might otherwise cause panic if disseminated to "the crowd."

Exhausting Volunteers

I had the privilege of attending the 2011 Humanitarian Action Summit in Cambridge, Massachusetts, that was organized by the Harvard Humanitarian Initiative, a university-wide research program supported by the Provost and the Harvard School of Public Health. One of the working groups of the summit explored the implications of using crowdsourcing approaches for disaster relief, crisis management, and violence prevention.

Those participants who were veterans of crowdsourcing enterprises using Ushahidi's platform for the response to the earthquake in Haiti were at that time working on a digital map to show locations of violence during Libya's pro-democratization struggle. A team of crisis mappers had been asked by the UN's Office for the Coordination of Humanitarian Affairs (OCHA) to create and post the map as soon as possible. The site went live during the summit, on March 5, 2011.[29]

Over the next day I spoke with a number of Ushahidi staff members and volunteers. I learned that while there is a perception that crowdsourcing makes information management easy, the reality is different. Often, location data are difficult to get. Cell phone service providers do not always provide information on the location of where a message is sent. Although the service providers could determine location by triangulating distances from three transmission towers, they do not always do so owing to concerns over confidentiality. So if a text message does not state a specific location, volunteers and staff members can spend a considerable amount of time trying to place the event. Following the earthquake in Haiti, however, people living outside of Haiti—especially friends and family members of Haitians—assisted in identifying locations from text messages.

In the aftermath of disasters, location tracking can be traumatic. One exhausted volunteer explained how she spent many sleepless nights trying to hunt down locations in Haiti after getting a text message from someone pleading for help. In one case, the person had texted a location with a name that corresponded to three locations in Port-au-Prince. The volunteer explained how she could not go to bed until the woman, who could have died, was found. This was one of many such incidents.

This kind of stress and exhaustion cannot be sustained. It can be minimized, however, if trust networks, perhaps with some paid staff members, are deployed during disasters. They can then train people in reporting about events and can distribute GIS devices to obtain exact locations. It is not a lot of money in the grand scheme of disaster response or violence prevention efforts to pay for Global Positioning System (GPS) receivers for volunteers who already have cell phones and attend a short training program.

Measuring Results

Ushahidi's maps have received mixed reviews. In response to the earthquake in Haiti in 2010, these categories were plotted on a digital map: emergencies, vital lines, public health, security threats, infrastructure damage, natural hazards, services available, and other. Currion analyzed data generated following the earthquake relative to their usefulness to those providing humanitarian response, and concludes:

Harsh as it sounds, my conclusion was that the data that crowdsourcing of this type is capable of collecting in a large-scale disaster response is operationally useless. The reason for this has nothing to do with Ushahidi, or the way that the system was implemented, but with the very nature of crowdsourcing itself. . . . The data that this crowd can provide is unreliable for operational purposes for three reasons. First, you can't know how many people will contribute their information, a self-selection bias that will skew an operational response. Second, the information that they do provide must be checked—not because affected populations may be lying, but because people in the immediate aftermath of a large-scale disaster do not necessarily know all that they specifically need or may not provide complete information. Third, the data is by nature extremely transitory, out-of-date as soon as it's posted on the map. Taken together, these three mean that aid agencies are going to have to carry out exactly the same needs assessments that they would have anyway—in which case, what use was that information in the first place?[30]

In fact, as Currion points out in his blog post, Ushahidi is recognizing that crowdsourcing can be of limited usefulness. As Chris Blow, a Ushahidi developer notes, "One way to solve this [problem of large amounts of unusable data]: forget about crowdsourcing. Unless you want to do a huge outreach campaign, design your system to be used by just a few people. Start with the assumption that you are not going to get a single report from anyone who is not on your payroll. You can do a lot with just a few dedicated reporters who are pushing reports into the system, curating and aggregating sources."[31]

One use of the Ushahidi platform for crisis mapping is in data processing. Currion points out that "to have available a recognised group of data processors who can do the legwork that is essential but time-consuming would be a real asset to the community—but there we've moved away from the crowd again."[32]

A distinction should be made regarding users. Humanitarian organizations may well find Ushahidi's maps to be of little value since aid workers typically provide bulk commodities (such as rice), potable water, hygiene packs (containing soap with which to prevent scabies, for instance), and other nonfood items (such as pans) to large groups of people. They have many workers "on the ground" and know where water bladders have been set up, where food is being distributed, where clinics are available, and so on. So a digital map that reflects information on needs and resources might be of limited value. In contrast, for individuals needing food, water, soap, pans, or clinical care, seeing what is available at the nearest location is extremely valuable. In the case of Haiti, it may well have been life-saving for quite a few people. Even though Internet access was limited, people called their relatives and friends in other parts of the country or in other countries who clicked into the detail of the Ushahidi map at specific locations, finding helpful information. In brief, while some humanitarian organizations may not have found the maps to be especially helpful, individuals at a local level did. This mirrors, once again, the transition from international outsiders to people "on the ground," but in this case, they were assisted by people all over the world—a great example of the undercurrent described in the preface. The undercurrent flow from *international outsiders* to *people at a local level* was complemented by *international insiders and those helping in solidarity*—not large humanitarian organizations, but individuals. As explained by Robert Munro, who was in Haiti during the humanitarian response, where he saw people using Ushahidi's site:

The biggest users of this system were not humanitarians, they were Haitians. A trickle of messages ended up on the public Ushahidi map, a river went through the entire system, but these pale in comparison to the absolute flood of messages between the crisis-affected population and their friends and relatives outside of . . . Haiti. Members of the Haitian community were using the map when in contact with their friends and relatives within Haiti that only possessed cell phones. They were directing people to the nearest locations [where] they could obtain food and explaining the system for obtaining food (eg: "there is a food distribution point 1 kilometer north, and you can only collect food for so many people"). The volunteers I was working with did this for many of the people who texted to 4636 [the Ushahidi number], too—it was a community helping itself and this was an order of magnitude greater than anything we in the humanitarian world achieved.[33]

In contrast to crisis mapping, we are just now beginning to learn about the usefulness of a platform like Ushahidi for violence prevention. It is an emerging technology. We do know that data from crowdsourcing is potentially complementary to other data, such as those derived from situation reports sent from trained and trusted field staff members or volunteers in a trust network who are capable of determining what is rumor and what is not. We also know that the increased situational awareness afforded by crisis mapping can help prevent violence in disasters that can occur due to delays in material support. In other words, to the extent that equitable distribution of aid is facilitated during disasters, and a perception of fairness is maintained, riots that would otherwise erupt can be minimized.

In Haiti, very few reports (constituting 1.5 percent) were made relating to security and potential violence. In two instances, emergency response was provided that arguably curtailed likely riots. On January 17 and February 2, 2010, violence was prevented by dispersing crowds that had formed in a plea for more food (Heinzelman and Waters 2010, 10).

While Ushahidi's technology is arguably the most frequently used by organizations using a crowdsourcing approach, a number of others have been and are being developed (by, among others, the ICT for Peace Foundation and IRIN Humanitarian News and Analysis (2010), a service of the UN Office for the Coordination of Humanitarian Affairs.[34] Engaging volunteers in reporting about crises and potential violence will improve our early warning capacities. My sense, however, is that there is no substitute for supporting local capacity so that early warnings will result in early response. It is human nature to want to communicate about something traumatic that is happening or is likely to happen. But it is counter to human nature to put oneself in harm's way.

It is clear that one advantage of crowdsourcing platforms is a dramatic reduction in cost. It is also reasonable to assume that the overwhelming thrust of crowdsourcing is on the warning side, not the response side. Yet the inexpensiveness of the approach frees up money for building local capacity for early response, or developing trust networks using a bounded crowdsourcing and bounded crowd-feeding approach, with restricted feeding for warnings.

We would be remiss, however, to overlook the impact that using social media and platforms like Ushahidi can have. These technologies make it possible to have "smart crowds," led by those who have a strategy, committed to bringing about major change.[35] In January 2011, Egypt had a group of activists, inspired by the transformation in Tunisia, who had studied strategic nonviolence, apparently inspired by the writings of Gene

Sharp, particularly his book *From Dictatorship to Democracy* (2003), as well as Mahatma Gandhi, and Martin Luther King Jr. The Egyptian activists communicated with thousands of protesters to overthrow the regime of then President Hosni Mubarak. This was not the case in Libya. The protest there turned into a civil war. Erica Chenoweth succinctly clarifies the differences, pointing out that "oppressive regimes need the loyalty of their personnel to carry out their orders. Violent resistance tends to reinforce that loyalty, while [nonviolent] civil resistance undermines it. When security forces refuse orders to, say, fire on peaceful protesters, regimes must accommodate the opposition or give up power—precisely what happened in Egypt" (2011, A27).

How effective are such protest movements? That is a matter of considerable debate. According to Chenoweth, she and Maria Stephan "compared the outcomes of hundreds of violent insurgencies with those of major nonviolent resistance campaigns from 1900 to 2006; we found that over 50 percent of the nonviolent movements succeeded, compared with about 25 percent of the violent insurgencies" (Chenoweth 2011, A27; see also Chenoweth and Stephan 2011). Of course, this is quasi-experimental because the sample is not randomized and there could be a self-selection process at play. In other words, nonviolent strategies might be more apt to be applied in "easier" cases. But most social science is quasi-experimental in that it is not possible when testing hypotheses with human subjects to subject one group, for instance, to violence and another to nonviolence.

In situations of strategy at a macrolevel, Ushahidi's technology can be used by organizations to map what is happening, not unlike what one sees in World War II movies, where generals smoke cigars and pipes and push what appear to be toy tanks, planes, and ships to different locations on huge map tables as their strategy evolves over time. Ushahidi provides a digital version of such a map table, but it works even better in combining textual information with pictures and videos, along with an ability to communicate quickly with large groups of people who submitted the information posted onto the map—those in the "crowd," members of a trust network, or both.

We are only beginning to understand how platforms like Ushahidi's will impact the world. But such platforms used with other technologies might prove instrumental in making a cluster of technologies work in ways that are "greater than the sum of the parts," to use a systems perspective.[36]

As explained in chapter 1, violence prevention at a local level can have a strategic impact, depending upon the stage of a conflict. In a pre–massive

violence phase, combining it with nonviolent methods for political transformation (in an attempt to keep troublemakers from ruining an opportunity) can make it possible for societies to address genuine grievances
without substantial bloodshed, thereby preventing an escalation to massive
violence. Similarly, these technologies can be used to prevent violence in
a post-massive violence phase thereby neutralizing the impact of spoilers,
decreasing the chances that a society will revert to massive violence again.

To the extent that these technologies (social media, crowdsourcing,
bounded crowdsourcing, digital mapping, crowd feeding, bounded crowd
feeding, and restricted feeding), used together, can increase the chances
that nonviolent social movements will have greater success, then they
collectively turn the warning-response gap on its head. Preventing violence
at a local level can make it possible to bring about change at a macrolevel.
There is no magic here, nor do I want to overstate the potential of the
combined use of these technologies and by doing so mistakenly imply a
technological panacea. More than eight hundred people died in Egypt
during the overthrow of Mubarak. Success in this arena involves people
taking the initiative at considerable risk. And even when dictators are
overthrown, preventing violence remains tenuous. As Emad El-Din Shahin
writes:

The transitions in [Tunisia and Egypt] have been far from smooth. In Tunisia, the
election of the Constituent Assembly, initially scheduled to take place this July, has
been postponed to October to give political actors more time to prepare for writing
a new constitution. Egypt has been experiencing similar calls for delay. Some groups
are advocating a "constitution first" process, contrary to the results of the referendum that took place in March, and are urging the postponement of legislative elections scheduled for September.

In both countries, the structures of the old regime have not been completely
dismantled. An atmosphere of mistrust among political elites, particularly between
liberals and Islamists, is precluding the emergence of needed consensus on key
issues. Major differences exist on such issues as the nature of the state, the role of
religion in society, and guidelines for creating a new democratic system.[37]

Summary

The availability of GIS data has increased the potential for greater precision
in early warning and early response at a local level. Georeferencing events
data creates the potential to identify violence or potential violence at a
specific location. Ushahidi is a leader of using crowdsourcing for early
warning at a local level.

One of the concerns about this crowdsourcing approach is that text messages and Internet postings are sometimes manipulative. They can be designed by troublemakers and spoilers to cause panic or concern. A similar problem exists, however, with other web-based approaches like Wikipedia. People can and do post erroneous, even libelous, content on a subject. Such postings misguide users and spread false information. Hence, a quality control mechanism of sorts is required.

Since time is often of the essence in conflict early response, checking data accuracy quickly is essential. Ushahidi has numerous approaches to determine data accuracy. One approach is to appeal to morality—appealing to *crowdsourcing honesty*. Another is mathematical, involving a sort of "peer review" process—referred to as *crowdsourcing the filter*. Other measures include assessing the reputation for accuracy of reports from the person sending it, the extent of redundancy in reporting of the same event, and proximity of the reporter to the event.

More "low-tech" approaches to checking validity of data are to send a message back or to call those who submit reports that seem dubious, asking for more information, or communicating with members of trust networks, asking them to look into whether an event happened. Creating and relying on a trusted group of reporters is called *bounded crowdsourcing*.

We know little about the relationship between reporting about actual or potential violence and intervening to prevent violence. Two potential ways to increase mobilization for violence prevention resulting from participation in a crowdsourcing process are to include more people in decision making and to mobilize through groups. In addition, organizations can use *bounded crowd feeding* and *restricted feeding*. The former involves sending back information to a core group of trained people who can intervene to prevent violence along with members of a crowd in close proximity to the potential outbreak. The latter involves sending warnings only to select individuals, mainly trained volunteers and staff members, along with trusted officials, if applicable, primarily due to concerns that warnings can cause panic and hysteria.

Ushahidi's crisis mapping technology was of considerable value to people at a local level following the earthquake in Haiti. This is an emerging technology; we are in the early stage of assessing how it might be used by organizations focused on conflict early warning and early response. We do know that data from crowdsourcing potentially complements other data, such as those derived from reports sent from trained and trusted field staff members who are capable of determining what is rumor and what is not. We also know that the increased situational awareness afforded by crisis mapping can help minimize violence in disasters that can occur due to delays in material support.

It is clear that the main advantage of crowdsourcing platforms is a dramatic reduction in cost and the potential to communicate with thousands of people with ease. Those within a trust network can be provided GPS receivers so that accurate location data are secured, reducing the prospect of exhausting volunteers.

It is also reasonable to assume that the overwhelming thrust of crowdsourcing is on the warning side, not the response side. Nevertheless, bounded crowdsourcing, with a trust network of trained people in selected locations, combined with bounded crowd feeding and restricted feeding, can potentially provide the leadership necessary to engage people in constructive collective action.

7 Circumventing Tribal Violence in East Africa

The few early successes of the [Conflict Early Warning—CEWARN] program [in the Horn of Africa] did not come from the highly developed technical side but from the very personal reporting side of those individuals gathering information before it was subjected to systematic extrapolations. Nevertheless, the systematic framework allowed the observer to ask the right questions and look for the data that revealed an impending crisis.

—Howard Adelman in blog posting of Patrick Meier, "A Conversation on Early Warning with Howard Adelman"

In 2007, I met with CEWARN staff members in Addis Ababa, Ethiopia, to learn more about their work. I was struck by how different CEWARN was from FCE. CEWARN is an entity of the Inter-Governmental Authority on Development (IGAD), a regional body with seven member states: Djibouti, Eritrea, Ethiopia, Kenya, Somalia, Sudan, and Uganda. I inferred as I listened to CEWARN staff members that being an official regional organization had its advantages and challenges. Advantages consist of having access to top-level decision makers and being able to secure governmental funding. Challenges include pressure to not report anticipated or actual violent incidents because of political sensitivities and related difficulties in navigating regional relative to national political imperatives. What is helpful to know for the vast majority of the region might be embarrassing to a specific member country. My sense, as I left Addis Ababa, was that CEWARN had very talented, hardworking, and well-meaning staff members, but they suffered from a "political straitjacket" that was both crippling and frustrating.

This chapter explains how CEWARN collects and analyzes events data, and is starting to use mathematical pattern recognition. It also covers how CEWARN is developing new technologies, including the use of high-frequency (HF) radios, high-gain antenna radios, community radios, and radio-frequency identification (RFID)—but these technologies are in early stages of implementation and their results, including costs and benefits, have not been analyzed thoroughly yet. Finally, chapter 7 reports results of CEWARN's program.

Background

The Horn of Africa has suffered from violence in remote rural areas for years. This is due, in part, to increases in population, pressures over scarcity of land, and environmental challenges. Violence is between different groups with different tribal identities and sometimes different ways of making a living. Some tribes are predominantly pastoralists (herders). Others are mainly agriculturalists (farmers). More recently, organized crime has become a factor. Highly lethal weapons are given to one tribe to raid the livestock of another. The crime organizations then sell the livestock at a profit. In an effort to mitigate this violence, CEWARN was created, with a mandate to track conflict and cooperation, and to issue warnings of imminent violence to member states.[1]

Starting in July 2003, CEWARN staff members in the field began submitting reports to their headquarters in Addis Ababa. *Situation reports* are produced weekly that convey information about the general tenor of relations (both of conflict and cooperation) based on answers to a questionnaire aimed at identifying conditions that could cause pastoral conflict. Situation reports sometimes provide warnings of likely violence. In contrast, staff members submit *incident reports* to document episodes of violence (Schmeidl and Mwaûra 2002).

CEWARN designed its events data categories on the basis of answers to questions that local experts identified as most significant in explaining when something is most likely to cause violence or facilitate cooperation. The categories are tailor-made to the conflict in the Horn of Africa. The reports of conflict and cooperation are developed with data, as explained in chapter 2, that reflects the *"who did what to who, where, when, and how"* of an event. Events derived from situation reports include these major categories: alliance formation, armed intervention, behavioral aggravators, environmental pressure, exchange behavior, mitigating behavior, peace initiatives, and triggering behavior. Events from *incident reports* fall into these major categories: armed clashes, raids, protest demonstrations, and other crime. The subcategories for each of these are found in appendix D.

For various reasons, CEWARN's focus has been on three IGAD clusters. Specifically, CEWARN has concentrated on the Karamoja (covering the cross-border areas of Ethiopia, Kenya, Sudan, and Uganda), Somali (covering the cross-border areas of Ethiopia, Kenya, and Somalia), and Afar-Issa clusters (covering the cross-border areas of Djibouti and Ethiopia).[2]

CEWARN has built a regional network of governmental and non-governmental organizations. This network contains four entities. First are Conflict Early Warning and Early Response Units (CEWERUs). These are national organizations comprised of both governmental and non-governmental leaders. Second are national research institutes (NRIs). These are independent contractors that collect and analyze information received from the field. Third are civil society organizations (CSOs), which serve a function similar to peace committees established by St. Xavier's and FCE. And fourth are CEWARN's trained field monitors (FMs). Through this network, CEWARN has the capacity to engage in violence prevention at national, regional, and local levels.

Financial support for CEWARN comes from IGAD member states and from international donors, including the German Agency for Technical Cooperation (GTZ) and the United States Agency for International Development (USAID).

CEWARN has a Rapid Response Fund (RRF) used to support projects in member countries. Grants of up to $50,000 are awarded for capacity building of peace committees, dialogue initiatives, and immediate support to vulnerable groups.[3]

Increasing Emphasis on Local Organizations

CEWARN staff members have developed extensive inventories (which they call "maps," since they list organizations by geographic location) of CSOs, which are either community-based organizations or NGOs, in specific clusters. As such, CEWARN is moving away from an organizational design that relies heavily on intervention by mid- and top-level leaders, in the direction of supporting local organizations. As a summary report of the Ugandan portion of the Karamoja cluster concludes:

There are positive trends in CSO capacities such as skills in CPMR [conflict prevention, management, and resolution], office infrastructure, grassroot[s] mobilization, networking, fundraising, lobbying and advocacy skills to mention but a few . . . which if harnessed by CEWARN could further improve the outcome if well coordinated and harmonized in a manner that optimally utilizes the available resources and capacities. . . . CSOs and DPCs [district peace committees] should position themselves to benefit from CEWARN's RRF, ICT for peace and the pool of trainers in CPMR trained by CEWARN. (Muhumuza and Bataringaya 2009, 11)

CEWARN periodically issues timely alerts. Here is a verbatim example of one that was issued August 4, 2010:

Conflict Alert

From the Ethiopian side of the Somali Cluster—Dire Woreda

Field reports are emerging that Borana Community settlers in El Dintu, Sololo District Kenyan side are attempting to re-route trade goods that are destined for Forole, Chalbi District, Kenya. The Borana settlers had initially been in Ethiopia for the use of grazing land, upon migrating back to El Dintu along with Ethiopian Borana community members information has linked . . . [Borana settlers] with attempts to control the trade flow into Forole.

The targeted trade route extends from Turbi to Forole (Kenya) and is used for transportation of many goods. Since the residents of Forole are primarily Gabbra, this attempt to thwart trade may expand to communal dimensions and raise tensions. Reports have also indicated that the Gabbra community on both the Ethiopian and Kenyan side are mobilizing and preparing to attack if the trade flow is disrupted. There is a high possibility that the conflict may expand to a more large scale confrontation between Gabbra and Borana communities. This could derail the highly successful implementation of the Dukana/Maikona Peace Accord. Local authorities and specially the district peace committees are advised to take the necessary steps to prevent any eruption of violence and possibly secure the trade route until both communities cease mobilizing for attack. **END**

Report Source: Tsegaye Bekele, CEWARN field monitor, Dire Woreda

Information Source: Waqoh Guttu, Dire Woreda security officer on the Ethiopian side of the Somali cluster

Over a period of eight years since its founding in 2002, CEWARN issued eleven of these alerts publically.[4] Of course, it is reasonable to assume that other alerts were issued privately. But eleven alerts over eight years is an average of 1.4 per year. This is indicative of what I referred to at the beginning of this chapter as a "political straitjacket"—alerts can be embarrassing to governments. But that is precisely why CEWARN's more recent emphasis on CSOs is so noteworthy. Local capacity is being built and alerts are being communicated verbally at a local level, effectively bypassing those straitjackets. At the same time, CEWARN is able to engage top-level leaders due, in part, to its intergovernmental composition.

According to Herbert Wulf and Tobias Debiel, CEWARN's outreach to top- and mid-level and local leaders has "managed to build confidence and collaboration amongst various stakeholders including governments, civil society organizations, and community-based organizations." At the same time, they argue that CEWARN still lacks an effective response component (Wulf and Debiel 2009, 19).

Technological Innovation

In recent years, CEWARN has developed digital mapping, SMS alerts, and a statistical module that allows analysts to generate forecasts. In addition, with the support of Virtual Research Associates (VRA), CEWARN is combining an analysis of fundamental indicators—such as infant mortality rates and openness to trade—to complement its use of events data. Long-term forecasts (one to fifteen years out) are added to analytical reports in addition to shorter-term forecasts and warnings. The long-term analyses provide a measure of "resilience" relative to more recent trends. Technical indicators are analyzed using market tools like moving average convergence divergence (MACD), adapted from those used by stock traders, an approach first developed by Schrodt and Gerner (2000).[5]

CEWARN field officers are expected to produce situation reports on a weekly basis, and to provide incident reports whenever there has been a failure to prevent violence. The reports include instructions of what needs to be done at CEWARN headquarters, such as making an urgent phone call to a government official. More recently, field officers are asked to complete *response reports* that cover what was done following a warning, who did what, and whether the intervention resulted in success or failure. These reports, over time, will provide CEWARN with a better understanding of how to make its programming more effective.

To help assess data reliability, CEWARN staff analyze field reports to determine whether redundant questions are being answered differently, not unlike approaches used by public opinion pollsters. In addition, the data are weighted based on the numbers determined through calibration focus groups comprised of local people who determine the potency of different forms of cooperation and conflict.[6]

A major challenge, however, in using technology for violence prevention in some developing countries is the lack of access to cell phone towers, especially when the violence being addressed is in remote locations. CEWARN staff members found that some violence-prone areas in the Karamoja cluster were up to 400 kilometers away from telecommunication coverage. This limitation, of course, compromises the effectiveness of a strategy that relies on staff members in the field to notify government officials when violence seems likely—not just in the restriction on communication, but also in the practical limitation of physical proximity for timely intervention.[7]

After spotting this communications deficiency, CEWARN hired a consultant from Kenya who launched the ICT4Peace project (not to be confused with the ICT4Peace Foundation in Geneva). With his help, CEWARN managers decided to use HF radios and high-gain antenna phones. HF radios operate at frequencies between 3 and 30 megahertz. They are referred to commonly as "ham radios." These can be used in places where there are virtually no cell phone signals. High-gain antenna phones, in contrast, use a combination of an antenna and a bidirectional amplifier. This also fosters communication where cell phone signals are weak or nonexistent.

CEWARN has also used community radio as a way of fostering communications within and between people of different identities. In doing so, CEWARN seeks to change "deep-rooted cultural beliefs and attitudes that contribute to the current cycle of conflict. These dialogues . . . also help communities in exploring alternative ways for economic gain and prestige other than cattle rustling."[8]

Community radio entails the use of a transmitter for communication at a local level. This added technology is indicative of a shift in CEWARN's strategy. Not only will mid- and top-level leaders be contacted when violence seems likely, but the communities themselves are being asked to get involved. This is similar to CeaseFire's approach in changing norms and community mobilization.

According to CEWARN, the communication technologies being used and the peace committees that have been formed have had a positive impact on violence prevention. According to Muriuki Mureithi, CEWARN's ICT consultant, "Numerous success stories of effective response interventions based on timely information received through the radios to mitigate potential conflict in Uganda have been reported. In Kenya, the radios have been used to transmit early warning information on an impending health crisis. The main challenge so far in implementing the project has been delay—in some instances—of securing licenses and frequencies for operation of the HF radios."[9] (It is important to recognize, however, that external evaluations of these approaches have not yet been conducted.)

In addition to using radio communications, CEWARN is developing livestock traceability capacity. This technology uses radio-frequency identification (RFID) microchips. Some use a battery; others do not. These microchips can be read to determine ownership of an animal.

The effectiveness of using RFID with cattle as a deterrent to rustling has not been researched extensively. Siror et al. (2009) propose an approach, however, that looks promising. It involves using tags that are difficult to see and hard to remove (such as those implanted under the skin), reporting to a central registry, and issuing alerts when movement out of normal grazing patterns become evident.

In Botswana, a reticular bolus RFID has been used with positive results. It involves a ceramic-coated microchip that the animal swallows. It is relatively cheap. According to a Department of Animal Health official, "initially, farmers could fight over the ownership of stolen and recovered cattle. But with the introduction of digital ID for cattle, the police with the help of field extension officers have recovered thousands of stolen livestock and identified their rightful owners. Incidences where two farmers dispute the ownership of a cow only for the police to find out that none of them is the real owner of the cow have also abounded." The department claims "at least" a 60 percent reduction in cattle theft. Keep in mind, however, that this requires a digital ID system, cooperation between countries where cross-border operations exist, and cooperation of slaughterhouses and police.[10] Whether these conditions can be met in the Horn of Africa, and with what impact, are open questions.

Results

Much of CEWARN's technology has not been evaluated thoroughly. Clearly, however, CEWARN's approaches merit further study because violence in remote areas in developing countries pose difficult communication challenges.

We do have evidence that CEWARN's field monitors—working with mid- and top-level leaders, along with CSOs—have been successful in preventing violence at a local level. My research assistant and I analyzed CEWARN events data and found statistically significant evidence that violence mitigation *reduces* organized livestock raids, deriving a negative correlation of 0.209 (Bock 2009, and see appendix E). This finding differs from that derived by Meier, Bond, and Bond (2007) who, surprisingly, found a statistically significant result that mitigation *increases* organized raids. Our opposite finding was due to our use of a different method that controls for the overall increase in violence over time. The two findings can be interpreted to mean that violence mitigation is effective at a local level, but it does not necessarily have an impact on violence on the wider societal or regional level. As discussed in chapter 1, violence prevention is most critical, from the standpoint of peacebuilding *strategy*, at the early and late stages of massive violence, while of value *intrinsically* and *tactically* at other times.

Summary

CEWARN is an entity of the Inter-Governmental Authority on Development (IGAD), a regional body with seven member states, and therefore provides us with an example of a conflict early warning and early response enterprise that is different than each of the other organizations covered in chapters 3–6. While the NGOs described in those earlier chapters face political challenges, they are not impacted by regular governmental involvement. This chapter, therefore, provides us with a view of both advantages and disadvantages of efforts to prevent violence by regional, intergovernmental bodies.

As with other violence prevention organizations, CEWARN is using text messaging and digital mapping to track cooperation and conflict in selected clusters of the region. Due to a lack of cell phone coverage on remote locations, however, CEWARN uses HF radio for communication with field staff. In addition, CEWARN is developing approaches to prevent cattle rustling that involve the use of RFID.

Over the years, CEWARN has developed more elaborate programming in support of local violence prevention efforts. This reflects the undercurrent mentioned in the preface that all levels of political, civic, and religious leadership are being engaged.

8   Comparing the Approaches

This chapter compares the costs and benefits of the different approaches. It offers propositions relating to peripheral vision, information processing, inductive reasoning, situations when there are only a small number of events, and the importance of timeliness in warnings and responses.

Costs and Benefits

When considering the costs of early warning and early response systems for violence prevention, it is helpful to keep in mind the major steps involved in setting one up. The steps needed with a "low-tech" approach are as follows: (1) form local organizations (such as peace committees); (2) gather information from trusted leaders in each community and build relationships at all levels; (3) train local organizations in violence prevention and engage community leaders as needed; (4) communicate regularly with peace committee members and various community leaders to keep abreast of any important developments and be prepared to intervene when tensions are high; and (5) advocate for justice so that the root causes of the conflict are addressed.

With a "medium-tech" computerized events-data system, these additional steps are needed: (1) develop categories of salient events, both for conflict and cooperation; (2) assign weights to events categories; (3) install software for events data storage and analysis; (4) install automated data generation software (if that is the approach one wants to use, though this is restricted to events reported in English-only media unless translation software is available and used); (5) recruit and train staff members for an information center in headquarters, and program staff both in headquarters and the field; (6) develop a security protocol for staff members working in situations of high tension; (7) create a standardized method of communicating trends to leaders at different levels and, if desired, to the general public; (8) if desired, add mapping capacity with which to depict cooperation and conflict trends; and (9) decide who should receive warnings and ensure that they have a way to be notified quickly (such as by sending text messages).

With a "high-tech" computerized events-data system using pattern recognition, in addition to the preceding activities, these steps must be taken: (1) hire staff members, or engage consultants, with high-level mathematical and computer science expertise; (2) design or acquire software with which to plot events and geographic data; and (3) determine what conditions need to be met for issuing a warning.

Finally, with another "high-tech" computerized message-aggregation system like Ushahidi's, these steps would be required: (1) recruit a core cadre of committed volunteers or staff members, or both, for the headquarters who have expertise in digital mapping; (2) download open-source software for the website; (3) promote ways in which people can contribute to the project by sending in reports via text messages, social media, or filling out forms on the Internet; (4) validate data; (5) post aggregated results on a map on the Internet; (6) send warning messages to selected people when incipient or actual violence is indicated; and (7) contact those who have an ability to intervene to prevent violence. Of course, in places where cell phones do not operate, radio technology can be used for communication and for tracking assets, like livestock.

It is useful to compare the costs and benefits of each type of approach in terms of their respective financial and human resource requirements, functionalities, and likelihood of sustaining themselves over time. Peace committees can, of course, endure from year to year, especially when officers are elected and there is an institutionalization of functions. Establishing them is relatively inexpensive or can cost nothing (if the community developers are volunteers). One can argue also that creating an automated system to generate events data inexpensively would have high set-up costs but could be relatively inexpensive thereafter (a sequence attractive to many donors who dislike making long commitments). A media company, such as a major newspaper, might be willing to host it. Such an approach could conceivably be sustainable. Similarly, an aggregation of messages, using the approach employed by Ushahidi, would involve some initial set-up expenses, but would be relatively self-sustaining thereafter, not requiring substantial funding over the long run, provided that the events data from messages have GIS coordinates and there is an automated method of posting them, crowdsourcing validity rankings. This might not be possible now, but it is clearly within technological reach.

To their detriment, "low-tech" approaches are not likely to be as diligent about information gathering when compared to others. Their data collection is typically passive, and communication with NGO field officers and leaders of peace committees tends to be unsystematic, lacking a perspective that those using a categorization scheme (like IDEA) enjoy. In addition, top- and mid-level leaders do not get the benefit of quantitative reports. And, in my experience, these types of systems are less likely to receive funding without the "bells and whistles" of other alternatives.

In contrast, with aggregated message systems like Ushahidi, it is likely that people will want to report information during tumultuous periods. Participating with others to profile an unfolding conflict where one can check the results periodically on the Internet raises the exciting prospect of contributing to something positive when one's world is seemingly falling apart. This kind of collaboration has the cachet of a social movement with members in regular communication, with the advantage of dynamic graphics. Therefore, it is arguable that these kinds of systems are sustainable over time to the extent that there is a dedicated group of staff members or volunteers, or both, to manage the website and to promote the contribution of messages. The role of NGOs with this kind of system can be to undertake data analysis and reporting (including sending warnings). There is no reason why these cannot be accompanied by training in violence prevention and tactical evacuation (as will be explained further in chapter 10), neither of which would be very expensive. If such systems function over long periods, time-frame adjustments will sharpen analyses by enabling users to ignore irrelevant or overwhelming quantities of data. But holding data for purposes of analyses over varied stretches of time is helpful, and might require that data be held in a cloud owing to its increasing volume as time goes by (which would have an added advantage of data security—an important factor when governments, militant groups, or organized crime syndicates see a system as threatening).

As explained earlier, the automation of events data generation can reduce costs significantly. Because information is derived exclusively from public sources, transparency is enhanced and security risks to field officers are lowered. The software must be purchased and currently is capable of generating events data only in English. And this type of information tends not to be geographically precise.

Even if they are more expensive, it is important to keep in mind that human-coded early warning systems fill gaps missed by automated systems. Humans are usually more able to provide geographically specific information (Bond et al. 2003, 8–9). Moreover, the discipline of information collection, buttressed by computerized pattern recognition, can make a difference by being instrumental in the issuance of a warning earlier in a lull than would be possible without such discipline.

When human and automated coding are combined, arguably greater efficiencies are attained than with purely human systems. Information gaps are filled by human coders, including more precise location information. At the same time, this approach takes advantage of markedly more efficient automated generation of events data.

From a financial perspective, among the five systems St. Xavier's and Ushahidi are, prima facie, the least expensive. St. Xavier's approach integrates early warning and early response into its ongoing humanitarian programs. And it builds local capacity. Ushahidi simply creates a website and posts messages it receives via SMS, social media, and Internet forms. The use of radios by CEWARN is a more expensive undertaking. The ability to communicate with others about a potential attack, or to reach out peacefully to another group through community radio, can be instrumental in saving lives, so the benefits of having ways to communicate in "cell tower deserts" can be substantial. It must be acknowledged, however, that hiring field officers to radio in information on conflict and cooperation is expensive, as is the case with FCE. Not only does it require hiring and training staff members in the field to collect data, and staff members in a central location to validate it, enter it, and analyze it, but also staff members need to intervene personally or else support the efforts of CBO members to intervene to prevent violence. And radio use can mean staff members are required at headquarters to review various media sources (in multiple languages) for events, code and validate them, enter them into a database, and write and disseminate reports and warnings.

Peripheral Vision: "Seeing the Forest Despite the Trees"

A medical definition of "peripheral vision" is: "Side vision. The ability to see objects and movement outside of the direct line of vision. Peripheral vision is the work of the rods, nerve cells located largely outside the macula (the center) of the retina. The rods are also responsible for night vision and low-light vision but are insensitive to color. As opposed to central vision."[1] In the case of early warning systems that pull together events data (whether from human coders, automation, text messages, social media feeds, or Internet forms), peripheral vision illuminates the big picture, revealing what might not be directly observable to individual peace committee members or field monitors. This illumination can (1) bring greater accuracy to reports and warnings due to active, rather than passive, events data collection; (2) identify the "holes" of relationships geographically (such as gaps concerning which moderate leaders should be called in a specific location if tension increases); and (3) produce analytical reports that add both time and space dimensions within a complex environment.

In the business world, those involved in direct sales or production need information to make decisions quickly, but they do not necessarily need to focus on the bigger picture. The managers, in comparison, do need to focus on the bigger picture, but would get lost in more disaggregated information.

FCE generated reports and Ushahidi produces maps that pull together disparate information into relatively coherent visualizations. Aggregate indicators (as in the number of killings) within selected locations and profiled in reports on a timeline are helpful to mid- and top-level leaders inundated by continuous floods of information. The reports bring some semblance of order to a complex, often confusing, information-saturated environment. This benefit has not been an integral part of either St. Xavier's violence prevention efforts or (until recently) CeaseFire's approach.

Another benefit of computerized events data approaches is that analysts are able to picture conflict and cooperation more broadly than simply at a district or precinct level. This advantage was mentioned by those working with FCE in their headquarters and users of FCE's reports. In the case of the Eastern Province of Sri Lanka, staff members sometimes spotted a "domino effect" of conflict spreading from one district to another.

Events data added together and depicted over time without pattern recognition can keep mid- and top-level leaders attuned to the overall picture. This can be helpful in encouraging these leaders to be more attentive and willing to help respond during periods of high tension. So the first proposition is: *While reports generated from events data that simply list the magnitude of events over time without pattern recognition are of little, if any, use to field staff members, they are valued by mid- and top-level leaders. To the extent that more of their attention is focused on simmering tensions as a result of such reports, their involvement in early response is more likely.*

Information Processing

During interviews with FCE field officers, they commented repeatedly about how the categorization scheme of the Early Analysis of Tensions and Fact-finding (known by its German acronym FAST) and the training they received from swisspeace and others on how to use the system made them think in new ways. Not only did they look more closely at various forms of conflict, they also began to be more attentive to signs of cooperation. Senior FCE staff members in Ampara District indicated that their training in data gathering and mediation methods was being incorporated into their daily work and specific interventions. Staff members constantly assessed issues of conflict and cooperation, though the database system was not always mentioned per se.

It is arguable that the discipline of collecting data can be instrumental in fostering interventions earlier during a lull than without a computerized early warning system. From a funding standpoint, to the extent that donors consider reports to be of value, as was the case with FCE's reports in Sri Lanka and CEWARN's in the Horn of Africa, proposing to build an early warning and early response system can assist in fundraising. And, if done well, an early warning and early response system can provide data for assessing impact of various programs.

Actively collecting information probably would not be pursued in such a disciplined way unless the data gathered by field officers is used for a valid purpose, namely for reports that will be disseminated widely. Imagine asking field staff members to turn in information at the end of every day because you want evidence that they are working actively in the field. Compare that to the need for information for a report that will help leverage intervention from the middle- and top-levels when they need it. Even though FCE's field staff members repeatedly commented that they saw no benefit from the data in terms of generating early warnings (due to the lack of pattern recognition capacity), the active collection of data means that field staff members are more likely to be informed of impending violence in order to initiate an intervention before violence occurs. And there is ample evidence that human induction of peace committee members, field staff members, and headquarters-based staff members yielded useful early warning decisions. This leads to the second proposition: *While reports that are generated from events data that contain trend analyses are helpful for middle- and top-level leaders, the discipline of active information collection, not the reports that are generated from those data, matters most at the local level, especially in enhancing inductive reasoning, even when mathematical pattern recognition is not being used.*

Inductive Reasoning and the "Small-n Problem"

One reason collective human induction is sometimes more useful than mathematical pattern recognition is that, in many conflicts, events that precipitate violent conflict occur infrequently at a local level. For instance, a rape that is perceived as having been politically motivated can invoke a violent reaction by those of the same identity as the rape victim, even when the crime occurs only once. This is not so much a question of a pattern as it is a single instance of a highly symbolic and disgusting event.

When there are small numbers of events used for pattern recognition, the "small-n problem" (that is, the small number of observations) can be overcome by expanding the geographical size of analyses.[2] And there are clearly some events that can be weighed heavily so that one or two instances of them can set off a warning. But this is not pattern recognition per se. To identify patterns, of course, other data—such as those accounting for hate speech by religious leaders, derogatory comments by a politician, hostile communications in newspapers, and provocative blog posts— would need to be factored into an analysis.

Another reason mathematical pattern recognition can be inferior to collective human induction is that often early warning and early response can be better served by the so-called wisdom of the crowds. That is, the collective wisdom of a group of concerned and frequently communicating people, close to the area of conflict, can make a more accurate assessment about violence than can a computer using events data and pattern recognition software. According to James Surowiecki, there are four elements required to make accurate assessments as a group: independence, decentralization, diversity of opinion, and aggregation (there must be some way for individual viewpoints to result in a collective decision—in this case, a decision to issue a warning, to intervene, or to evacuate). In contrast, dysfunctional collective assessments can result when there is homogeneity, centralization, impeded information flow, one viewpoint leading to other similar viewpoints ("cascading"), and high levels of emotional engagement (Surowiecki 2004, 66–83; Sunstein 2006).[3] Indeed, as Wigger points out: "Discrimination and bias can distort the collection and the analysis of information. . . . Communication with a community through a tool, such as phones or internet may well privilege the voices of those who are communication literate and have the tools available and discriminate against other groups that are less fortunate, such as women in particular contexts, the elderly, handicapped, and others" (2011, 19).

There are at least three distortions of crowd-sourced data acquired through social media. First, given that social media is used typically by younger people, crowdsourcing approaches will tend to have an age bias. Second, some evidence suggests that unrepresentative political philosophies are another distortion. And third, online information does not always distinguish what is coming from outside from what is coming from inside a situation. During riots in Tehran, Iran, in 2009, expatriates from Britain and the United States offered "more radical views than the home population directly affected by repressive measures" (Gujer 2011, 24).

Recall that Lawrence stressed the importance of FCE field officers and peace committee members communicating with those of other organizations locally (Bock, Lawrence and Gaasbeek 2009, 226). This information flow can benefit from the "wisdom of the crowds." At the same time, computerized pattern recognition, which can be designed to automatically produce warnings to selected people at all three leadership levels, can be a check on unhelpful group process dynamics. As Wulf and Debiel point out, "Human beings tend to assume that causes and effects stem from similar categories. They are thus not prepared to react appropriately towards accidental or unintentional effects . . . [they] tend to neglect the extent of negative events because 'moral feelings' do not work with numbers. A mass catastrophe may even result in inactiveness (psychic numbing)" (2009, 27). This leads to a third proposition: *Automated pattern recognition can be useful if for no other reason than being a hedge against pathologies in collective information processing.*

One way to combine pattern recognition with human induction can be seen in a Philadelphia Police Department initiative in 2008 and 2009. The department installed mobile data computers (MDCs) in its patrol cars. With this device, police officers are able to undertake *data mining* (searching data and using algorithms to identify patterns in it) and analysis in their vehicles as well as in their offices, using a tool called the "Crime Spike Detector," which is now called "Hunchlab."[4] Hunchlab sends out automated warnings to police officers via text messages. But the value of this approach, it seems, is greatest for individual officers in the field when they can query the system and get real-time results. And to the extent that officers gain knowledge over time that increases their analytical precision, learning leads to a more effective discernment of what to query and what parameters to use. Given that the Philadelphia Police Department's crime database has over thirteen million events and another five thousand-plus are added daily, there are plenty of data to use artificial intelligence-enhanced approaches, such as hidden Markov models. But this brings us back to our initial quandary, which is a fourth proposition: *We need to learn more about what computers can do that the human mind, either individually or collectively, cannot do as well in identifying patterns and anticipating events.*

A Premium on Timeliness

If Horowitz's contention is correct that rumors about impending violence are apt to occur toward the end of a lull, then low-tech systems that mobilize when rumors are detected will tend to produce alerts when there is not much time remaining before violence. But such warnings late during the lull are not especially detrimental to effective early response to the extent that these systems rely on people locally where the trouble is. People will, of course, hear things that are not reported in the news or on a website. They will probably hear "rumblings" even before they hear rumors. Computerized early warning systems have the potential to capture information not covered by the media if they have staff members, peace committee members, or volunteers in trouble spots who report what they hear and see (and, sometimes, "feel"). If information about rumblings is entered into an events database, the computer system's ability to issue a warning earlier than a low-tech system is a function of adequate data and capacity for pattern recognition. The question, from the standpoint of issuing a warning, is whether one way is superior to another in producing and communicating actionable information, and most importantly, to people at a local level—the likely victims of violence who have the greatest stake in its prevention and are most capable to facilitate its mitigation.

There is an inherent tension between the need for accurate information and the timeliness of getting it. St. Xavier's trained peace committee members to get to the bottom of rumors. This can be done relatively quickly by checking the facts of a rumor either in person or with a phone call. FCE had dedicated staff members for verification. These Colombo-based retired police officers often called active police in the locations where the events reportedly occurred. This can cause a delay of a few minutes, a few hours, or even days. In contrast, Ushahidi's approach to verification is relatively quick. Using crowdsourcing for "peer review" and using mathematical approaches involving measures of reliability, redundancy, and proximity to events do not take much time. And phone calls can be made expeditiously if needed.

Ushahidi also uses an approach similar to St. Xavier's to check the accuracy of rumors. Meier found that validity checks by staff members were conducted in a matter of minutes. When they receive an alert (a message that violence is about to happen which, if valid, would result in a warning being issued), they can "validate the information with any available reports from the news media. The team also can contact the individual who reported the alert to ask for further details on the event. . . . If someone sends an alert and subsequently gets a call from the Ushahidi...[a] few questions, asking for specific details, will make it apparent whether or not the person is lying."[5] Indeed, checking the validity of data is facilitated substantially with text message reports since the person has a cell phone and Ushahidi has the number. This, unfortunately, is not the case with data derived from social media or Internet forms.

At a meeting sponsored by Internews (an international NGO that fosters independent media and access to information worldwide) entitled "Improving Humanitarian Information for Affected Communities," Meier introduced Ushahidi's approach. In his blog entry about the meeting, he commented on the tradeoff between accuracy and timeliness of information:

The more we demand fully accurate information, the longer the data validation process typically takes and thus the more likely the information will become useless. Our public health colleagues who work in emergency medicine know this only too well.

The figure below [figure 8.1 that appeared in his blog entry] represents the **perishable nature of crisis information**. Data validation makes sense during time-periods A and B. Continuing to carry out data validation beyond time B may be beneficial to us, but hardly to crisis-affected communities. We may very well have the luxury of time. Not so for at-risk communities.

This point often gets overlooked when anxieties around inaccurate information surface. Of course we need to insure that information we produce or [on which we] rely is as accurate as possible. . . . Yes, we can focus all our efforts on disseminating facts, but are those facts communicated after time-period B . . . really useful to crisis-affected communities?

. . . At CEWARN we included "Source of Information" for each incident report. A field reporter could select from several choices: (1) direct observation, (2) media, and (3) rumor. This gave us a three-point weighted-scale that could be used in subsequent analysis.[6]

Figure 8.1 shows the time-relevancy relationship only. The basic idea is that the more time spent verifying the information, the less relevant it is for at-risk communities. When I showed this figure to Michael Clark, statistical consultant of the Center for Social Research at the University of Notre Dame, he suggested adding to it a depiction of utility and accuracy. In figure 8.2, relevancy is changed to utility and accuracy is given a separate axis. Time versus utility and time versus accuracy are both shown. The point of disutility is noted, but keep in mind that this phenomenon is highly contextual. In some situations one might start out with easily and quickly verifiable and thus accurate information. In other contexts,

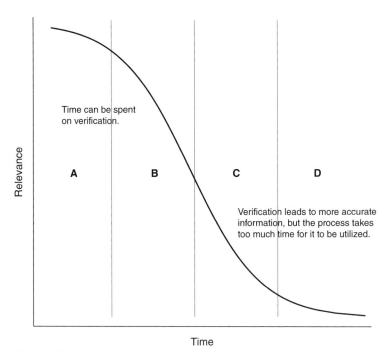

Figure 8.1
Time-relevancy relationship. Used with permission from Patrick Meier.

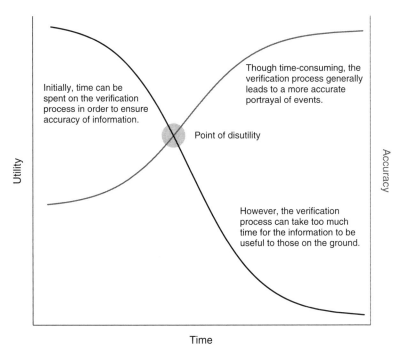

Figure 8.2
Time-utility-accuracy relationship

verification takes much longer. Sometimes the point of disutility might not be where more accurate information comes at too great a time cost (i.e., the cross), but instead is hard-coded to a specific time (e.g., fifteen minutes or fewer in the case of a heart attack after a 911 call in which no verification takes place).

There are, of course, no universally applicable answers about data accuracy compared to its timely use. But Meier's graphic depiction in figure 8.1, and the point he makes about early warning information, is to early warning and response what the *theory of second best* is to economics (Lipsey and Lancaster 1956). Sometimes, it is necessary to make a "second-best" decision when one does not have complete information. It is a suboptimal solution, but it is the best solution in view of suboptimal conditions. Meier's argument in favor of an assumption of validity is defendable if for no other reason than the fact that many challenges in politics and bargaining involve the use of "imperfect information." When a rumor of a planned attack is reported, every effort can be made to find out whether the rumor is valid. Sometimes there will be a preponderance of indicators that "point in the same direction," making the determination of validity plausible. In other instances, when additional indicators are lacking, a validity check can be made with a quick inquiry via telephone or in person, or by checking media reports. This leads to a fifth proposition: *Events data should be validated as a standard practice, but if the validation process lingers with no indication of foul play by one or more reporters, then a warning should be issued to select individuals when data indicate that violence will soon happen.*

9 How to Intervene Effectively

The idea of early warning, conflict prevention and preventative diplomacy is becoming widely accepted as compelling—even in vogue, one might say . . . [but] very little is known that has been done in actually translating it into operating strategies or workable practice.

—Éva Blénesi, "Ethnic Early Warning Systems and Conflict Prevention"

People are not averaging machines in which positive and negative information is given equal weight; instead, bad dominates good in the same way that a cockroach makes a glass of juice undrinkable. . . . we believe that negativity bias is an important part of the explanation of the relative weakness of peace building and the relative power of incitements to violence. Once recognized, this problem leads to explicit investment of resources in preventing or countering the negative intergroup events that incite violence. Preparations for rumor busting, mobilizing moral authority against violence, encouraging individual rather than group attributions for wrongdoing, and even building relationships with the police—require resources and attention to the extent that intergroup relations are dominated by negative events.

—Clark McCauley and Joseph G. Bock, "Why Does Violence Trump Peace Building?"

There are, of course, many ways to intervene to prevent violence. Much more research is needed about how interventions can be most effective in differing conflict situations. Here, I present examples that illuminate approaches relating to the analytical framework and theories in chapter 1. The examples are divided into these groupings: reducing emotional escalation and counteracting justification for violence, changing norms, reducing the sense of threat, increasing the sense of risk of becoming violent, and developing options of what to do if faced with violence.

Reducing Emotional Escalation and Counteracting Justification for Violence

One often hears the words "mob psychology" when reference is made to groups that have been violent. There are "football hoodlums" and ransacking crowds. And there is a sense that these groups become violent on the spur of a moment. Indeed, that can happen, but usually there has been some degree of "preparation" prior to the violence. And then, when an event occurs that invokes an emotional reaction, those who have been "stoking the fires" (to use the imagery provided by Brass 1997) seize the opportunity during a period of volatility. That is, those who have something to gain from violence usually do not have much control over a precipitating event. But they do have an ability to seize an opportunity for violence when a precipitating event occurs. In the case of football hoodlums, for instance, the troublemakers in the crowd have no control over whether a referee will make what is perceived to be an unfair call. But if a referee does so, they can seize the opportunity to lead people in a violent direction—presumably for the mere "fun and excitement" of a fight.

Where tight identity group cohesion exists along religious lines, intrafaith interventions are recommended, as explained in chapter 1. For example, FCE staff members and Co-Existence Committee members were instrumental in preventing an attack in Muttur in December 2005. They learned that Hindu Tamil youth in Ralkuli planned to attack Muslims, reportedly to avenge earlier violence against Tamils. FCE staff members and Co-Existence Committee members communicated with Tamil civic leaders, requesting that they intervene in an attempt to calm emotions and talk the youth out of attacking. This proved to be a successful intervention (Bock, Lawrence and Gaasbeek, 2009).

An example of taking away religious justification for violence relates to a plan to overtake the al-Aqsa Mosque in Old City of Jerusalem. This mosque is considered the third most holy site in Islam, after Mecca and Medina in Saudi Arabia, due to the belief that the prophet Mohammad visited there (some believe metaphysically, others figuratively) and ascended to heaven, called the *Al-Mi'rage*. The al-Aqsa Mosque coincidentally sits atop the remains of the temple built by the Jewish King Solomon. The surrounding wall of the temple, usually called the "Temple Mount" (but sometimes referred to as "Mount Moriah") is, of course, sacred to Jews.

When I worked in Jerusalem with Catholic Relief Services, we developed a project in partnership with the Harry S Truman Institute of Hebrew University and the Wi'am Palestinian Conflict Resolution Center that engaged Christian, Jewish, and Muslim students in researching religious incitement

(Bock 2001b). The students found a startling amount of venomous material on the Internet. One site, in particular, caught our attention. It was developed and updated by a group of young Orthodox Jews. The focus was on how to create a situation whereby Palestinians would become violent toward Jews at the Temple Mount, invoking a takeover of the al-Aqsa Mosque by the Israeli police and military. This, to them, would be a step toward heralding the coming of the Messiah (whom Jews believe has yet to come).

One of these groups planned to throw the head of a pig into al-Aqsa Mosque during Ramadan, the month when Muslims undertake the spiritual discipline of fasting. Learning of the plan of the Orthodox youth, Sephardi Chief Rabbi Eliahu Bakshi-Doron wrote to Yasser Arafat, chairman of the Palestinian Authority. His letter was quoted by Wohlgelernter and Rudge in *The Jerusalem Post*: "We were sad to hear of the criminal plot by extremists who wished to harm the faith and faithful and inflame relations between the religions. We denounce any attempt and evil thought which could put off peace and friendship" (1997, 1 and 6). In addition, the rabbi offered warm greetings to Islamic believers during their holy month of fasting. This letter was publicized and served to diffuse Jewish-Muslim tensions to the point that a violent incident did not erupt at the Temple Mount (Atallah 1998). Rabbi Bakshi-Doron discredited the religious zealotry of his coreligionists.

FCE reported another example of moderate religious leaders preventing violence. It occurred over a period of seven days, from July 31 to August 6, 2007. The report states that FCE

mediated to prevent a multi-party conflict among Sinhalese and Muslims, [and] Muslims and Tamils . . . in the Muttur and Seruvila divisions. This conflict had arisen after the [presumably Buddhist] Government Agent in Trincomalee had prohibited quarry work on the rock at 64 Mile post in the Muttur division on the information that there had been ruins of a Buddhist shrine on the top of the rock which was of archeological value. This was perceived by the Muslims and Tamils as an attempt to encroach land in Muttur by the Sinhalese Buddhists. Meanwhile, FCE gathered information that Sinhalese spoiler groups were attempting to establish a new Buddhist temple on this rock in front of Muslim and Tamil villages. Rumors also started spreading that the Buddhists in Sruvila [were] . . . planning to challenge the Muslims' and Tamils' identity by erecting a temple on the rock.

The inter-religious Co-Existence Committee in the Trincomalee District entered into a negotiation. This was witnessed by the Heads of Police and the Army and the Divisional Secretariat. A fact finding visit was undertaken by the religious dignitaries to the top of the rock. Everybody declared that they saw some ruins of archeological value only at one place of the rock. The Archeological department was called to demarcate the ruins. The Divisional Secretary allowed the rest of the area of the rock for quarry work. The Buddhist Priest of the Seruwila Temple declared that no new erection of Temples will be allowed on the rock. The potential outbreak of violence was prevented.[1]

And here is another example, from Afghanistan. According to Mary Anderson, who had spoken with a staff member of Norwegian Church Aid:

Two commanders and their mujahedeen ["religious warriors"] were preparing to battle over some incident that caused a rift and men were actually taking up positions in a village area to begin the battle. A mullah from the local mosque took out his loud speaker and ran up and down the streets proclaiming that "no one will come to the funerals of anyone who dies in this conflict." This signaled his judgment that the upcoming battle had no religious justification and that people who died would not, therefore, be martyrs (eligible for heaven). The battle did not occur. (1999, 29)

An alternative way of counteracting the sense that violence is justified is to "contaminate target purity," thereby making potential perpetrators feel that innocent people (quite possibly, their "own" people) will be harmed through the violence. When people become violent toward others there is a dichotomous mindset. It is "us versus them," a sense that "we are the good guys and they are the bad guys." In ethnic riots, for instance, Horowitz argues that in-group members go to agonizing lengths to specify precisely who is part of an out-group. Detailed lists are developed of who lives or works where. The importance of this is seen in debates and trials of identity that sometimes occur during the violence. Those whose identity is unclear are allowed to live (Horowitz 2001).[2]

Part of the reason for this excruciatingly careful attention to the purity of a target relates to risk, in addition to the determination of whether or not the killing would be "just." To the extent that members of the in-group feel they are unlikely to face consequences for their violence because the authorities will "look the other way," there is a possibility of prosecution if "innocent people" are killed.

A simple way to contaminate target purity is to send members of an in-group among members of the out-group that is about to be attacked, with in-group members showing resolve that they are prepared to be harmed or to die along with members of the out-group. As we saw with St. Xavier's, some in-group members showed resolve not to move and to "get in the way" of an attack while sitting on the front steps of the homes of out-group members. This was effective in preventing an attack.

Another way of reducing or eliminating a sense that violence is justified when that justification is based at least in part on religious beliefs is to engage in hermeneutic dialogue. In the heat of tension, however, in cases where religious beliefs are being manipulated to incite violence against an out-group, theological considerations are not always presented. I have conducted three different simulations with graduate students in which we divide the class into two different identity groups and designate moderate and extremist leaders in each. Prior to the simulations, the students work in small groups to develop their theological arguments, anticipating what the other will say about a scenario given to them of an impending attack. The students spend hours combing through various sacred texts, like a debate team preparing for competition. In each instance, while I antici- pated eloquent theological dialogue and debate, the moderates quickly moved into an appeal for civic virtue. The extremists, in contrast, tended to stick with their religious messages. While this is anecdotal, it suggests to me that both religious and civic messages, not religious messages alone, are likely to be used when moderates engage with militants of their own faith in a public setting in an effort to discredit extreme positions, thereby reducing the number of (usually spontaneous, not "hard core") followers of militant leaders.

Changing Norms

Recall that CeaseFire focuses on the modification of norms as one compo- nent of its theory of change. Essentially, CeaseFire seeks to eliminate any sense that violence is acceptable, or that living in a violent place is "just part of life." An example of changing norms comes from British India, in a place called Khyber Pukhoonkhawa, which is now contained within the North-West Frontier Province of Pakistan, where a culture that expected blood revenge, called the *badal code*, was challenged and modified success- fully. As Johansen reports:

After World War II ended and a postwar British government finally promised self- government for India, rioting broke out between Hindus and Muslims. [Mahatma] Gandhi and [Abdul] Ghaffar Khan traveled together trying to stop it, achieving mixed results at best. Ghaffar Khan returned to the North-West Frontier Province to address the problem there among his own people. Although the Khudai Khid- matgars [his group of followers otherwise called the "Servants of God"] movement had been explicitly, intensely Islamic, the leaders had also kept it non-communal [meaning that they did not foment xenophobic ethnoreligious nationalism]. One of its objectives had been the promotion of Hindu-Muslim unity. On one occasion, when Hindus in Ghaffar Khan's home region of Peshawar were threatened, his brother called in 10,000 Khudai Khidmatgars. All were Muslim, armed only with faith and courage, yet they were able to protect Hindus and Sikhs against rioters and to restore peace in the city. (1997, 64)

This was the first nonviolent army in the world. It was comprised of Pukhtoons, the same ethnic group that comprises the largest share of the Taliban. The nonviolent Servants of God were started by Abdul Ghaffar Khan, a close friend of Mahatma Gandhi. Together, they "fought" nonviolently against British colonialism (Easwaran 1984). This struggle also had the effect, albeit a temporary one, of changing the norm anticipating that those who have been harmed will seek violent revenge.

One reason Ghaffar Khan was successful in leading ethnic Pukhtoons was that he built on their identity, their history, and their sense of honor. The *badal* code demanded justice based on retribution, the exercise of which often required sacrifice and bravery. Being a Servant of God demanded a new kind of justice, a demand that came from Allah, and likewise required sacrifice and bravery. It involved standing in the way of potential attackers, putting one's life on the line in a different way, in an attempt to keep blood from spilling rather than spilling it. In so doing, Ghaffer Khan provided a viable option for ethnic Pukhtoons that was consistent with their identity and was based on their normative principles.

A grandfather of Ali Gohar, one of my former students who came from the North-West Frontier Province, was a member of the Servants of God. According to Gohar, his grandfather helped protect religious minorities during the partition of British India. He "fought" nonviolently alongside one of his formerly sworn enemies.

Gohar worked for thirteeen years with refugees in his hometown of Peshawar before becoming a Fulbright Scholar in the United States. When he completed his studies he returned home to found Just Peace International. His goal is to resuscitate the nonviolent movement among his people.

This is no easy task. As CeaseFire found, changing norms requires an intimate understanding of the culture within which that change is desired. In an interview with Gayathri Fernando, Gohar explains:

Honour and shame are an integral part of the Pukhtoons as they live in a community strongly linked through clan, sub-clan, tribe and sub-tribes. Each one is the custodian of personal and family tribe honour. Dishonouring one means dishonouring the whole tribe. An act of shame on the part of one will be considered an act of shame for the community of the same tribe. Honour is again divided in to *Gharat* and *Azat*. If a father, brother or relative is killed and the others cannot take revenge that is called *Gharat.* If, while a woman is walking with you and someone teases her or a guest is treated badly by some one [sic] in your *Hujra* (community centre), the person or community has taken your *Azat.* Such notions of honour lead to violence among individuals while other members of the clan and tribe take [a] stand, and the circle of violence will accelerate till elders intervene with an initiative of peace building. Regarding killing there are peaceful overtures that are possible. . . . One can easily imagine how honour is close to a single Pukhtoon. Claiming descent from the missing tribe of Moses, the Pukhtoons as children study in three different institutions mosque, school and *hujra*, the third being the community centre for each clan owned by the whole community where they learn all ethics and traditions of Azizwali (Pukhtoons code of life). Understanding the community culture and tradition is quintessential in order to change traditional attitudes to violence.[3]

Just as CeaseFire hires former gang members to be interveners, Gohar's Just Peace International works closely with ethnic Pukhtoons who understand the culture and are respected within it.

Reducing the Sense of Threat

As explained earlier, St. Xavier's used "myth busting" as a method to counter a sense of threat related to rumors. This is a time-sensitive undertaking that, if done well, can usurp incipient violence.

Another time when a sense of threat usually needs to be reduced to keep the peace is after violence occurs, when animosity and suspicion are endemic. My wife, Sue, and I researched and wrote a case study on how a group in Karachi, Pakistan, helped a massive slum area overcome this ubiquitous sense of threat. Here is an overview.

An NGO, Orangi Pilot Project (OPP) focuses on supporting people in the Orangi slum of Karachi, Pakistan's center of commerce. On December 14, 1986, there was a major riot in Karachi centered on Orangi. Immigrants from India have a different identity in Pakistan and are referred to generally as Mohajirs. After a Mohajir girl was killed by a bus driven by an indigenous Pakistani, of the Pukhtoon ethnicity, violence erupted between the Mohajirs and Pukhtoons. The results: 150 deaths, 320 damaged vehicles, 1,307 damaged and looted houses, and 1,768 damaged and looted businesses.

OPP responded quickly to the damages in Orangi, distributing food and clothing. They also worked with a partner NGO to support the rebuilding and replacement of houses. During reconstruction, a sense of threat was reduced by involving both Mohajirs and Pukhtoons in the process. According to OPP's founder, Dr. Akhter Hameed Khan, "The result was that not only was the economic cycle rehabilitated but also the attention of so many people was turned from vengeance to reconstruction" (Lyke and Bock 1995, 15). According to the findings of the Local Capacities for Peace Project (otherwise known as the "Do No Harm" project for which my wife and I wrote the case study on OPP), people in post-conflict settings often are able to work on common tasks, and doing so brings about the extent of reconciliation that is possible at that time (Bock and Anderson 1999). In situations in which "wounds are still open," focusing on a common task is far more effective than engaging people in intergroup dialogue. Their healing is not far along, and their sense of threat remains, albeit in a dwindling state.

Increasing the Sense of Risk of Becoming Violent

An example of increasing the sense of risk so that people are reluctant to participate in violence is provided by FCE. On July 7, 2005, I received a brief report from Madhawa Palihapitiya, director of programs at FCE. Here are verbatim excerpts from the message, with minor editorial changes:

On the 17th of June two unidentified . . . men shot dead a police sergeant attached to the National Intelligence Bureau (NIB). Several violent incidents took place in Serunuwara & Ali Oluwa areas in Trincomalee in the wake of [this] killing. . . . Several lorries transporting . . . [relief supplies] for Tsunami victims [were] attacked by Sinhala youth. Tamil civilians who were using the Ali Oluwa junction for traveling were also assaulted. A vehicle belonging to the Sri Lanka Monitoring Mission [SLMM] that went to the site in order to investigate the attack on a lorry was also attacked. Sri Lanka Army (SLA) personnel rushed to the site and provided security to two Scandinavian monitors . . . and the civilian driver. . . . The SLA escorted them to Trincomalee. The LTTE Trincomalee District Political Head . . . made a complaint to the SLMM Trincomalee regarding the assault on Tamil civilians [because the assault was a violation of the LTTE-GoSL ceasefire agreement].

On the 19th of June a Sinhalese home guard was abducted by a group of armed men suspected to be LTTE. On the 22nd of June a group of Sinhalese youth stopped a passenger bus traveling from Kantale to Muttur & assaulted 3 Tamil women. The assaulted women . . . made a complaint to the LTTE officers in Killivedidi immediately after the incident. After 2 hours a group of Sinhalese youth came under a grenade attack in Dehiwatte area in Trincomalee causing injuries to [three people].
. . .

A meeting [was] convened by the Trincomalee FCE [staff] on the 23rd and 24th of June at the Co-operative hall in Seruwila. . . . The discussion was held with the Sinhalese community leaders, on the 23rd of June 2005. They insisted on negotiations with the LTTE and the Tamil community leaders and agreed to hold a meeting with 100 community leaders from 10 villages & five Buddhist monks in the first phase. . . .

On 27th of June the mutilated dead body of the home guard abducted by an unknown armed group a week [earlier] . . . was recovered by a few villagers. . . .

On 28th of June, FCE Trincomalee District team headed by the District Coordinator . . . met the LTTE Trincomalee District Political Head . . . in Sampoor where [the LTTE Trincomalee District Political Head had] agreed to meet the Sinhalese Buddhist civil leaders to discuss the current political situation & asked the FCE to organize the meeting. On 1st of July 2005 the FCE took 15 Sinhala Buddhist Civil Society leaders from 10 Villages in Seruwila to LTTE-controlled Sampur where there was a successful meeting between the two parties which was facilitated by the FCE. Both parties agreed to resolve their disputes amicably through negotiations in the future. The FCE is now arranging a much larger mediation between the two parties with the Security Forces, Police, [and] government officials [with] the Sri Lanka Monitoring Mission participating as observers.

From this message we can see that Co-Existence Committee members were instrumental in being a "first line of intervention." They notified FCE Trincomalee staff members about the tense situation. It also appears that FCE Trincomalee quickly made contact with FCE Headquarters in Colombo. Finally, the Sri Lankan military's response was forceful and timely, and police and government officials were engaged in reducing tensions, all of which served to *increase the sense of risk* in becoming violent. While violence occurred, it did not break out into a massive incident, due to a host of actions by community leaders, the government, SLMM, and FCE.

Recall, as well, that St. Xavier's sought the help of police officers in Ahmedabad when violent passion erupted due to the building of a Hindu monument near the Shahpur Fire Brigade station on public property. The presence of the police made it clear that anyone who became violent would be arrested.

Developing Options for When Violence Threatens

Developing options in highly tense situations often requires creativity tempered by realism—what Lederach calls a "moral imagination." This imagination involves four components: (1) "the capacity to imagine ourselves in a web of relationships that include our enemies"; (2) "the ability to sustain a paradoxical curiosity that embraces complexity without reliance on dualistic polarity"; (3) "the fundamental belief in the pursuit of the creative act"; and (4) "the acceptance of the inherent risk of stepping into the mystery of the unknown that lies beyond the far too familiar landscape of violence" (Lederach 2005, 5).

A scene in the movie *Friendly Persuasion* illustrates such moral imagination during the U.S. Civil War. The movie is about a family of pacifists, members of the Society of Friends (also known as Quakers). The mother of the family, determined to live out her beliefs, refuses to leave the family homestead as belligerent troops enter the region. Her husband is absent, looking for their son who decided, not without trepidation in shunning his family's beliefs, to fight with neighbors to counter the attack.

As the mother sees smoke rising from destroyed farms, the burning progressively coming closer, she decides she will cook a meal for the enemy soldiers. As they approach, looting and shooting anyone deemed to be a threat, she surprises the soldiers by inviting them into her house and offering them a hot meal. The soldiers pause, stunned to be treated this way. They accept the offer and sit down and enjoy the hot meal. The woman, her daughter, and youngest son are not harmed. Why? It is hard to say: the soldiers were hungry and enjoyed the food, the woman did not treat them like enemies and did not show (at least substantially) that she feared them, and the encounter served to reduce the kind of demonization that is typically required for human beings to kill other human beings.

This, of course, is not a common scenario. Many efforts at accommodation, however creative, result in bloodshed. This is due in part to a lack of information. When we discuss conflict early warning at a local level, useful information is not solely about when there is likely to be an attack; it is also about the intentions and past behavior of the oncoming belligerents. As poignantly put by Casey Barrs:

Page one of any negotiation manual will stress: Know who you're dealing with. . . . Information gathering can reveal how belligerents have treated civilians elsewhere. . . . What do verification visits and survivor testimonials reveal? Did belligerent demands remain tolerable or become unbearable? Did they follow through more on their promises—or their threats? Which faction leader can actually enforce promises made? Skills that result in better situational awareness (*and* readiness to safely disengage if necessary) enable civilians to pursue accommodation to its full limit as opposed to breaking off prematurely (due, for example, to opting to fight based on manipulative provocations or to take flight based upon false rumors). . . . The better that local leaders can disseminate facts and develop plans, the better able they may be to keep their own people on a nonviolent footing. Anyone who feels that information gathering and preparedness can itself be destabilizing (which is true) must face the fact that it is often *mis*-information and vulnerability which is destabilizing. Populations caught between fighting factions are often fed information that is deliberately false and inflammatory. It is calculated to tempt or incite them to fight.

But local leaders . . . armed with facts and proofs may be better able to belie belligerents' fear and hate mongering propaganda.[4]

One of the most widely known approaches to developing options is *interest-based negotiation*. Popularized by Roger Fisher and William Ury in their book *Getting to Yes* (Fisher and Ury 1981), this approach involves helping groups in tension to identify their *interests* as compared to their *positions*. Researchers at the Harvard Negotiation Project found that those involved in negotiation typically tend to settle on their demands—in other words, their positions. But when people are helped to clarify their interests, there is a far greater range of possible solutions. This does not necessarily mean that everyone gets what they want. It means there is potential to identify more common ground than if people are frozen in adamant stances with specific positions.

This is explained well by Silke Hansen, a senior conciliation specialist of Community Relations Service:

My favorite illustration . . . is the story about the mom who comes in when the girls are fighting about the orange. Everybody thinks they know that story but they don't. She splits it in half and they each get half. After the mom does that she sees that one squeezes it for juice and the other grates the peel. So if the mom [had] recognized their interests, they both would have gotten all of what they wanted. A good mediator will understand the concept behind that but a great mediator will take the illustration one step further and say OK one girl wanted juice, but her need [wasn't] orange juice it was needing a beverage. She might have been happy with water or milk, or my personal favorite beer, or coffee or whatever. The other girl needed a seasoning. So if it wasn't orange, it could have been maple or vanilla. Any number of options there.

If you really talk and focus on what the needs are, the orange is one possible solution, but there are lots of others. Beginning mediators are so focused on the orange and who is going to get the orange that they lose out on a whole spectrum of other possibilities because they are allowing the [parties] to limit the discussion around who gets the orange and [which] parts of it. In fact the possibilities of resolving this conflict are much broader. We are not even getting into the possibility that the girls were not really fighting about the orange, but the orange was just a convenient object at that point because they fight over everything. I'm not even going to go there but that might be another piece of it. Regardless, if we just focus on the orange, we are limiting their possibilities, and we are limiting our abilities to help them deal with the conflict.[5]

One of the examples Lederach gives of offering another option while using "moral imagination" occurred in La India, along the Carare River (Rio Carare) in Colombia, South America (2005, 13–16). This river is located in a violent area of Magdalena Medio. There are armed groups that demand allegiance of poor farmers (otherwise known as "peasants" or *campesinos*). The river is used to transport both petroleum and drugs and their accompanying armed groups.

This part of Colombia became a battleground in the 1960s because the guerrilla Armed Revolutionary Front of Columbia (FARC) entered the area, followed by the Colombian military. Due to the ongoing violence, land owners decided to create their own private militias, often working in collaboration with government soldiers. These militias became infamous for breeding a violent environment of lawlessness and intimidation.

In 1987, a captain of the Colombian army spoke to a crowd of some two thousand farm families of La India and offered them forgiveness for their alleged allegiance to guerrilla forces. But, in exchange, he said they were expected to join his militia and fight the guerrillas. He gave them this choice: "You can arm yourself and join us, you can join the guerrillas, you can leave your homes, or you can die."

One of the farmers, however, Josué, did not like being boxed in by these limited options. Alejandro García, who researched this incident, quotes the speech that Josué made in responding to the captain:

You speak of forgiveness, but what do you have to forgive us? You are the ones who have violated. We have killed no one . . . you will not facilitate even the minimum credit for our farming needs. There are millions for war, but nothing for peace. How many men in arms are there in Colombia? By rough calculation, I would say at least 100,000, plus the police, plus 20,000 guerrillas, not to mention the Paras [i.e., the paramilitary soldiers], the drug lords and private armies. And what has all this served? What has it fixed? Nothing. In fact, Colombia is in the worst violence ever. We have arrived at the conclusion that weapons have not solved a thing and that there is not one reason to arm ourselves. We need farm credits, tools, tractors, trucks to make this little agricultural effort we try [to] make produce better. You as members of the National Army, instead of inciting us to kill each other should do your job according to the national Constitution, that is, you should defend the Colombian people. Look at all these people you brought here. We all know each other. And who are you? We know that some years ago you yourself were with [the] guerrilla[s]. . . . You brought people into our homes to accuse us, you lie, and you switch sides. And now you, a side switcher, you want us to follow your violent example. Captain, with all due respect, we do not plan to join your side, their side or any side. And we are not leaving this place. We are going to find our own solution. (García 1996, 189, as quoted by Lederach 2005, 14–15)

While it was one thing to express a desire to find another solution, it was quite another to build consensus about what that might be. A small group of farm leaders ultimately decided they would use civilian resistance, and would not use weapons. They formed the Association of Peasant Workers of Carrara (ATCC). And they developed six principles as their guide to the future:

1. Faced with Individualization: Solidarity.
2. Faced with the Law of Silence and Secrecy: Do everything publicly. Speak loud and never hide anything.
3. Faced with Fear: Sincerity and disposition to dialogue. We shall understand those who do not understand us.
4. Faced with Violence: Talk and negotiate with everyone. We do not have enemies.
5. Faced with Exclusion: Find support in others. Individually we are weak, but together we are strong.
6. Faced with the Need for a Strategy: Transparency. We will tell every armed group exactly what we have talked about with other armed groups. We will tell it all to the community (García 1996, 200, as quoted by Lederach 2005, 15).

Table 9.1

Component for violence	Interventions	Examples
Emotional engagement	Calming emotions	Prevention of an attack on Muttur, Sri Lanka, by asking Tamil civic leaders to calm Tamil youth, urging that they not attack Muslims.
Justification for killing	Coreligionist discredits potential perpetrators	Rabbi denouncing the plot to foster violence near the Temple Mount in Jerusalem.
Changing norms	Developing a culture where violence is frowned upon	Servants of God surrounding non-Muslims as a human shield in the North-West Frontier Province of Pakistan, substituting nonviolence for the *badal* code.
Reduction of risk of violence	Articulating consequences of being violent	A show of force by the Sri Lankan Army following a grenade attack on Sinhalese youth.
		CEO of Safari.com threatening prosecution if text messages were used to propagate hatred in post-election Kenya.
A feeling of no other options	Building support within a community	Creating a civilian resistance and safe haven option in Rio Carare, Colombia.

After the group developed these principles, representatives were sent to each local village. Eventually, farmers posted signs with a title that read "What the People Here Say." The signs then listed the principles and also noted that weapons were not allowed in the area.

Maintaining this as an unarmed sanctuary required meeting with each of the armed groups separately. These meetings were followed by a public debriefing and were not closed-door sessions. The farmers consistently maintained transparency.

This approach was successful in the area in which it was implemented (although Magdalena Medio has continued to be plagued by violence). In 1992, the farmers' group was recognized for its success by the United Nations. Sadly, however, Josué and several other leaders were assassinated, allegedly by hired guns (*sicarios*) working on behalf of local politicians, not armed groups. Those with "moral imaginations" paid a heavy price, but they were successful in identifying and implementing an alternative option.[6]

Summary

Some of the examples covered in this chapter are summarized in table 9.1. As emphasized previously, these examples relate to the analytical framework and theories presented in chapter 1. There are, of course, many other approaches that can be used.

10 What to Do When Violence Prevention Is Unlikely to Work

The victim-centric approach assumes that the communities and individuals at risk are themselves critical actors in the protection process. Protecting and promoting their rights, dignity and integrity is essential for the effectiveness of the work. They should play a key role, influence decisions, and make practical recommendations based on their intimate understanding of the nature of threat.

—Andreas Wigger, "Protection of Civilian Populations in Conflicts and Other Situations of Violence: The Challenges of Using ICTs"

Conflict early warning and early response mechanisms are usually geared towards prevention or mitigation of violence. The reasoning is that if it is known that there is a high possibility for emergence of violence and response strategies have been set out, then the concerned stakeholders can do something so that the conflict comes to an end or continues with non-violent means. This however does not always work and if it does, often comes too late. Another strategy, which is increasingly being applied in disaster management, would be to put more effort in helping to prepare the people who are at risk.

—Kristel Maasen, *Mobilising Early Response to Prevent Violent Conflict: An Overview of Obstacles to Early Response*

We should be concentrating our efforts on developing technology that will help communities to improve their own resilience, so that they have to rely less on external organizations to support them in times of crisis. . . . Unfortunately that isn't the model that we have right now, especially when it comes to technology; historically information has been extracted from affected communities. . . . Rarely do communities receive information back again in a useful form.

—Paul Currion, "Conclusion"

Early warning and early response approaches often have treated the local population as if they are passive recipients of informed, brave, and well-prepared interventions from the outside. This view is akin to a perspective about poverty alleviation, sometimes called the "white man's burden" (a gender-biased phrase). Putting the two together, one gets a perspective that reads: *These poor and vulnerable people need our help. If they are poor, we must provide charity by feeding and clothing them. If they are vulnerable, we must bravely step forward and intervene to protect them.*

An alternative view has been called people-centered early warning and early response. The most eloquent proponent of this approach is Casey Barrs, a former humanitarian worker in the Philippines, Somalia, Pakistan, and Zaire (now the Democratic Republic of the Congo) with American Refugee Committee and Mercy Corps, who researched this problem as a fellow at the Cuny Center (named after humanitarian worker Fred Cuny, who disappeared in Chechnya in 1995). Barrs claims that outsiders often suffer from well-intentioned humanitarian hubris. This is due, in part, to focusing mainly on early warning and early response at a macro- rather than a microlevel.

Barrs not only offers an insightful critique, but also suggests two solutions—*preparedness support*, on the one hand, and *locally led advance mobile aid* ("mobile aid" for short), on the other. Preparedness support is for situations in which conventional aid agencies have access to local populations—but perhaps not much longer due to deteriorating security (Barrs 2010). Its premise is to support local capacity for self-preservation in preparation for the possibility of when aid agencies evacuate due to insecurity or are otherwise unable to provide protection.[1]

Conversely, mobile aid is for situations in which conventional aid agencies are unable to gain access to endangered populations. Barrs argues that mobile aid cannot be conducted by conventional NGOs. It should be done by a new kind of NGO with the mindset and skill set to recruit, train, equip, and send local teams back into their home areas with specialized "survival" skills.

Preparedness Support

Preparedness support is conceived especially for use by international NGOs and transnational organizations working with internally displaced people (IDPs), refugees, and other vulnerable populations. It involves getting ready to evacuate or hibernate (that is, stay in a safe place until a violent period is over) when violence is imminent.

In the humanitarian field, much emphasis has been placed on disaster preparedness (see, for instance, International Federation of Red Cross and

Red Crescent Societies 2009). The focus is on training communities about what to do when there is a natural disaster. I witnessed such an approach when I worked for Catholic Relief Services (CRS) in Pakistan during a flood along the Chenab River near Multan. Months before the rainy season, Oxfam/Great Britain had trained villagers along the river how to respond to flooding. Villagers built grain silos on stilts to protect what for many was a year's supply of wheat and, on higher ground, prepositioned other food, chlorine tablets (for purifying water), and soap (for washing to prevent scabies). They knew where to flee when flooding was imminent and monitored flood warnings by transistor radio.

Because CRS was focused substantially on agricultural development projects, not humanitarian relief, our staff members had not worked with villagers prior to the flood. But because of the devastation, our work in flood response was still helpful in rescuing families trapped on rooftops with no clean water and very little food (which we did in collaboration with our partner NGO, Pattan). We were helpful in "mopping up the mess," but Oxfam had been instrumental in "minimizing the mess" for those communities it served. Oxfam's approach prevented loss of life, livestock, and precious food. Both interventions were helpful, but from the standpoint of minimizing human suffering and cost effectiveness, Oxfam's approach was superior. I mention this not to be critical of CRS's work in Pakistan. The CRS staff members there did excellent agricultural development work. Emergency response was not the programmatic emphasis. But this example nevertheless highlights the value of *preparedness* compared to *response*.

It is helpful to compare preparedness in the face of an impending flood with preparedness in the face of imminent violence. To stretch the analogy, rather than fighting a flood with sandbags and the like, it is sometimes better to flee and go to higher ground. That is to say, there are instances when the response should not be to engage with potential attackers but, rather, to leave. Barrs expresses concern that there is not enough emphasis on when to get "out of harm's way. So even micro level responses appear to follow the industry pattern: formal-legal-civil engagement with belligerent parties rather than informal-physical-tactical disengagement from belligerent parties."[2]

The nagging question that haunts Barrs and other humanitarian workers of international nongovernmental and transnational organizations is: "Have we helped prepare them [the local people] to face violence alone?" In a litany of concerns, he argues that our *"presence-based* and *program-based* ('mainstreamed') protections end when we leave. Our *camp-based* protections (tied to its physical layout and perimeter) become irrelevant when we or they are forced out of those refugee or IDP camps. . . . Our *rights-based* protections may be lost the minute we leave them to a lawless situation" (2010, 5).

In fact, Barrs goes a step further. In his experience, he has seen how international humanitarian organizations can inadvertently put people at greater risk, doing more harm than good, violating the Hippocratic oath of "do no harm," when they create the impression of security when there is none.[3] In a passionate email message, he writes:

The support we lend local organizers can . . . embolden them. But this might happen at the very time death lists of organizers are being drawn up. And it can cause them to delay preparations for their own safety, which they might otherwise have pursued had they not felt protected by our presence and imprimatur. Our accompaniment has its limits as demonstrated in the decimation of our protégé partners and civil societies of Haiti, El Salvador and elsewhere. The protective presence we offer can also be extremely shortsighted. We can be involved in day-to-day humanitarian maintenance for years without real strategic foresight as was recently demonstrated in Darfur, where expatriate capacity was "devastated" by expulsions. The expelled agencies had not spent any of the preceding 5 to 6 years bolstering their beneficiaries' capacity to be physically safe or economically survive alone in the face of violence.[4]

Barrs argues that humanitarian workers must embrace a code of ethics that recognizes this harm and implements measures to counteract it. We need to plan for a "remote aid vehicle" when we can no longer stay. Remote villagers, refugees, and IDPs living in camps must have accurate information as to whether they have protection from armed militias, and must be ready to flee (if negotiation is imprudent) when they are about to be attacked. And when they flee, they need to have prepositioned life support. They must stay away from obvious escape routes and disperse widely so they are not easily followed. They must know where to flee to safety, if such places exist (Barrs 2010).

There is precedent for supporting the tactical abilities of locals to stay physically safe.[5] The Darfur Peace Agreement tasked peacekeepers of the African Union Mission in Sudan (AMIS) in Darfur to create community police. Since then, local volunteers have been advised on communication, patrol skills, and security techniques. Likewise, the United Nations Mission to the Democratic Republic of the Congo (UN MONOC) helped to create a network of 524 villages in South Kivu, organizing patrols and rudimentary early warning mechanisms that sent alarms of pending attacks to peacekeepers deployed in the area.

Organizations are now becoming more knowledgeable about how to support local capacity in preparation for violence. As the United Nations High Commissioner for Refugees (UNHCR) says, "security mechanisms involving refugee guards, wardens, patrols, and watch teams can be highly effective" (UNHCR 2004, 79). This has been undertaken in Liberia, Sierra Leone, Guinea, Tanzania, and Ghana, and among Angolan, Congolese, and other refugee groups. As per the agreed upon safety package, aid has included equipment and telecommunications to accompany training in early warning, mediation, first aid, and safe movement. Field safety advisors with military or police backgrounds play a key role in "setting up refugee warden and community policing systems" (UNHCR, 32). Their help "aims to maintain the humanitarian and civilian character of the camps" that are facing armed attacks (Herrmann 2003, 8–9). UNHCR sees capacity building as a "softer" protective step on its "ladder of options." Some NGOs already "facilitate the development of community watch groups" that they equip with mobile phones and other communication equipment (O'Callaghan and Pantuliano 2007, 35; Women's Commission for Refugee Women and Children 2006, 22).

Barrs divides preparedness support into three components: physical safety, economic survival, and local service delivery. Briefly, physical safety includes training on information management, communications, safe sites, safe movement, and threat response. Economic survival includes guidance on (1) supporting local coping tactics (such as in developing lending networks and remittance schemes), (2) protecting assets (through dismantling and hiding, for instance), and (3) transferring assets (which has the double benefit of protecting family wealth while keeping that wealth out of the hands of criminals and belligerents). Local service delivery involves open service (that is readily visible) but also—when open work is no longer possible—deinstitutionalizing aid delivery to more discreet forms. An example is to send health workers from house to house rather than having a clinic in a village. A clinic supported by an INGO is visible to controlling powers.

Mobile Aid

Mobile aid involves the recruitment of small local teams. NGOs provide training and equipment to support these teams in providing services, preserving assets, and assisting, if need be, with hibernation (staying put, preferably out of sight of potential belligerents) or evacuation. The teams return to their home areas to help their communities.

There is a challenging lethargy that many communities suffer even when they face—and are aware that they face—danger. Barrs finds that to an astonishing extent, people in conflict die because they do not believe "rumors or exaggerations" of danger. They often simply do not want to believe that their government or fellow citizens are committing atrocities. Nor do they want to uproot their families and move their belongings. They need proof—but proof often comes too late, unfortunately along with their own entrapment. In the same way that international emergency response faces what Barrs calls a *paradox of early warning*, so too does local preparedness to evade violence: *the earlier a warning is, the less compelling a threat seems* (Barrs 2004).

Mobile aid teams may help arrange survivor testimonials for at-risk populations, or facilitate verification visits. These measures are not designed to create panic. The "healthy fear" that people gain about an approaching threat is channeled. The message is that danger is coming, but there are steps that can be taken. At mobile aid demonstration sites civilians can see what "conflict preparedness" might entail. They can see, for example, how to eke out an existence when they are dispersed and hidden. They can watch how to "dismantle" their village homes and build temporary huts. Or they can inspect a local early warning system. Demonstration sites can be located in areas on the verge of being destabilized, which are still relatively secure but have a palpable sense of threat. In such circumstances, people feel a "need to know" and are therefore more receptive to learning than when they feel secure.

Because "community lethargy" can endanger vulnerable communities when credible threats are escalating, assertive and credible leadership is required to facilitate an orderly evacuation. This can be provided by mobile aid teams working in tandem with local leaders.

I can relate to this type of collective lethargy. My family and I lived in Jerusalem from 1997 to 2000 when I worked for Catholic Relief Services. Saddam Hussein was ruling Iraq and, at one point, he threatened to launch SCUD missiles with biological or chemical warheads at Israel/Palestine. Because CRS was implementing projects funded by the United States Agency for International Development (USAID), it was granted permission to purchase "gas masks" and other sequestration items for staff members to create or keep in a "safe room" in their homes. We were instructed to identify a small space, use thick polyurethane sheeting and tape to prevent any leakage of air, store food and water, and become acquainted with using our masks and air canisters. (Our young children thought the odd-looking gas masks were great fun and would "model" them for visitors!)

My niece was living with us for the year. We started receiving phone calls from my brother and his wife, concerned about their daughter's safety. We eventually decided to send her home.

In the meantime, as the SCUD missile threats from Iraq continued, USAID came out with another communication which gave their grantees permission to use grant money to evacuate. The evacuation site identified was Cyprus. It was generally understood that those involved in projects benefiting Palestinians were expected to stay back, letting their families flee to safety.

This, of course, was an alarming communication. There was a great deal of conversation among various USAID grant recipients and contractors doing work in the West Bank and Gaza Strip about what to do. My wife and I agreed that I would stay back in Jerusalem, and that she and our two children would fly to Cyprus. The problem was that, according to U.S. government regulations, anyone evacuated with U.S. government grant money could not return until three months had passed. When we discussed this separation with our children, they expressed considerable sadness and concern.

My wife and I felt ambivalence about what to do for days that seemed like weeks. It was not until one Sunday after a church service that we made a firm decision about what to do. Many of the expatriates joined together after the service to have an open discussion about the dilemma we faced. People were concerned about what might happen, and also heartbroken about splitting up their families. Finally, one of our friends spoke up. He challenged us to look deeper into our faith. He said our faith should help us overcome our fear. While this might sound like some sort of end-of-the-world cult that one hears about on the news, the message he delivered spoke to our hearts. After that, my wife and I decided to stay in Jerusalem, to keep our family together. Of course, if the risk of an attack became too great, we could all flee together with our own financial resources, not the U.S. government's—assuming, of course, that we would be able to leave before it was too late.

Looking back, it is astounding to me how difficult it would have been for us to leave. We did not have an extended family there (although our network of friends seemed like one). We were renting an apartment; it was not a house passed down from generation to generation. Our belongings were meager, because we had been living in Pakistan before that and had brought very little with us. Almost all of our family pictures, our special dishes, and other sentimental items were packed safely in a storage shed in the United States. And we had a secure "safe haven" waiting for us. The

U.S. government would have flown us to Cyprus, all of us if we had wanted. It would have put us up in a nice hotel for three months.

How much more difficult it must be for vulnerable villagers to leave their homes and homelands. They have deep ties with friends and family. The homes that many of them live in were built with their own hands or those of their relatives. All their belongings are in that one place, and there is considerable uncertainty about what they will face if they flee.

It is understandable that it is difficult to uproot one's family and oneself in the face of violence. To do so seems quite unnatural. Therefore, it requires forceful leadership.

I am sure that many of us who were working for INGOs felt a sense of cognitive inconsistency at the thought of evacuating when those we were trying to support, and our dedicated local staff members, would be left behind. In that sense, we had an ethical dilemma not unlike that facing many international humanitarian workers. As Barrs explains:

A new skill set, new mind set, indeed a whole new aid architecture is needed for our local staff and partners who are left to face violence alone. We do not engineer much safety into our remote aid vehicle before pulling out. Until the day we leave, we still mentor them in Cadillac aid. We groom them to be good subcontractors with sizable "absorptive" and "surge" capacity. . . . [W]e try to model them after ourselves—even though our own highly evolved aid machinery, with all its administrative, logistical, and financial capacity, is not designed to survive conflict. It is unethical to make others responsible for an aid vehicle that has not been truly retrofitted for work amid violence. We should help prepare their *survival* abilities as much as their *service* abilities. (Barrs 2010, 23; my emphasis)

International humanitarian organizations and UN agencies, especially those working with refugees and IDPs, often work in unstable political and military environments. Many of them have security protocols for their staff members, including when to travel (usually only during the day and not on certain routes that are, for instance, infested with guerrilla roadblocks), the regularity with which to call headquarters by cell phone or radio, a standing directive to flee at any time one feels in danger, and a requirement to hibernate or evacuate when ordered to by a supervisor. They also usually have policies and procedures for the preservation of cash and major capital equipment such as vehicles and computers.

Most of these organizations assign people in their headquarters to monitor the security conditions wherever they have staff members throughout the world. It is not infrequent that staff members in the field feel a disinclination to evacuate when they are ordered to do so by supervisors in headquarters. (I personally have seen a staff member in the field tender her resignation after being given a choice of evacuating or leaving the organization.) Field workers typically do not want to evacuate because they feel they are not in danger, regardless of the security assessment of head-quarters' staff members. The people in headquarters look at information from numerous sources and changes in security conditions over time. There is a sense in which field staff members suffer from the "boiled frog syndrome." That is, if you put a frog into a pan of water and turn on heat beneath it, the frog will stay in the water until it dies because it does not recognize the gradual change in temperature. In contrast, if you drop a frog into boiling water it will jump out soon enough to save its life (Senge 1990, 22). So, the thinking goes, humanitarian workers in the field are often so close to the situation, and accustomed to persistent violence, that they have difficulty seeing the extent to which they are in danger.

The other reason that international NGO and UN staff members typi-cally resist evacuating is that they feel a commitment to their fellow staff members who are nationals of the host country or refugees who entered that country on their own. Humanitarian organizations typically will not take responsibility for evacuating those whom they have not sent to a specific country. This is because they do not have the authority to do so from the standpoint of emigration, not to mention the logistical difficulty of an evacuation involving considerably more people.

So imagine being in a war-torn country with fifty staff members, only six of whom are from outside the country. You receive an order to evacuate because there has been a warning of a violent uprising, pending the outcome of an election. You are asked to deposit whatever cash is in the safe into an international bank, to purchase plane tickets for all expatriate staff members and their dependants, and to leave as soon as possible. How would you feel if you had worked with the other forty-four staff members during the prior two years, if you had shared dinner in their homes with their families, gone to the weddings of their children, and celebrated holi-days together?

I recall agonizing discussions about these ethical challenges when I worked at American Refugee Committee. Usually, donors are not agreeable to letting national staff members (that is, people on the staff from that country) use vehicles acquired with grant money for "non-official" pur-poses. And it is very difficult to draw lines as to who can use those vehicles to flee. National staff members would, of course, like to transport their entire extended families. With a limited number of vehicles, who has prior-ity? It would not be right for only the top managers to get permission to flee in the organization's vehicles.

So, international humanitarian workers themselves have trouble evacuating. It is little wonder that they have difficulty supporting preparedness to evacuate vulnerable people they serve.

An Aversion to "Military-Style" Tactics

One of the arguments made by Barrs is that it is a glaring oversight for us not to help humanitarian aid beneficiaries, local staff members, and partners prepare for a failure to prevent violence. We have an ethical responsibility to prepare them in case an early response does not work, or it is not even prudent to try. Essentially, Barrs argues that early warning and early response systems, when taken to their logical conclusion, need to have a robust component of preparedness support. International NGOs and UN agencies need to support local, vulnerable populations' capacity for self-preservation when working with those populations even when a threat is not imminent. If there is no access to those populations, humanitarian organizations need to be prepared to support a mobile aid organization.

To embrace this approach, there is a sense in which humanitarian organizations will need to have a quasi-military capacity. On the one hand, Barrs admits that those working in humanitarian organizations "have visceral reactions" to anything involving the development of "skills that sound militaristic" (Barrs 2010, 12). On the other hand, humanitarian organizations' nascent work in "community policing," interventions to mitigate military threats such as landmines, and strategies for disaster risk reduction—which have many tactical and command-and-control elements—show that there are already programmatic initiatives in the direction of building quasi-military capacity at a local level. Perhaps calling it "survivalist training" is more palatable.

An example of terminology that tends to make NGO staff members uncomfortable is the name of a group engaged in preparedness support and mobile aid of IDPs in Burma (not referred to here as "Myanmar" because it is a preference of some who live there, including the Karen people, to refer to it as "Burma"). The organization is called the Free Burma Rangers (FBR). While in many contexts the word "ranger" denotes a helpful parks and recreation official, in the United States it is viewed by some as associated with Army Rangers, a highly trained group of commandos.

By the FBR's own description, its members are not engaged in military activities, but they do work under the protection of "ethnic resistance armies." Their purpose is to be a multi-ethnic humanitarian service movement. They bring help, hope and love to people in the war zones of Burma. Ethnic pro-democracy groups send teams to FBR to be trained, supplied and sent into the areas under attack to provide emergency medical care, shelter, food, clothing and human rights documentation. The teams also operate a communication and information network inside Burma that provides real time information from areas under attack.

In addition to relief and reporting, other results of the teams' actions are the development of leadership capacity, civil society and the strengthening of inter-ethnic unity. The teams are to avoid contact with the Burma Army and operate under the protection of the ethnic resistance armies. However, they cannot run away if the people they are helping cannot escape the Burma Army. Men and women of many ethnic groups and religions are part of FBR.[6]

FBR claims to have trained over 110 teams since 1997, 48 of which are working full-time in "the Karen, Karenni, Shan, Pa'O, Arakan, Kachin, Chin and Lahu areas of Burma. The teams have conducted over 350 humanitarian missions of 1–2 months into the war zones of Burma." Part of their work involves health care. They estimate they help between 1,000 and 2,000 people each mission.[7] For years locally recruited teams have been trained by the Free Burma Rangers to bring relief and security advice to threatened communities inside Burma. Two separate networks of discreet three- to seven-person teams have collectively served some 280,000 displaced persons. The Burmese government has seen these net-works as a threat. Consequently, some of the team members have been killed.[8]

I am not being critical of the name Free Burma Rangers. In fact, the organization's name supports the contention of Appleby who describes both violent and nonviolent fundamentalists as militants. He reserves the label "extremists" for violent militants (2001, 11). They use approaches not unlike "special forces" in helping people survive. That is why they appear to some to be "militant" in the sense that they use military tactics, and they are prepared to die (but not to kill).

I spoke with Dave Eubank, the founder and director of FBR. He is com-mitted to the point of risking his life and those of his family as a result of his faith.[9] He explained how Burma has been divided by the dictatorial government into three zones. White zones are under governmental control. Black zones are not, and have over one million displaced people in them. And brown zones are in between. He explained how he and FBR teams work in black zones nine months a year. He told me there is one commu-nity on average that is attacked every year by the Burmese military result-ing in community member and Free Burma Ranger casualties.[10]

Often, international NGOs and UN agencies ask for peacekeepers in situations where insecurity prevails. Humanitarian organizations tend to be comfortable with skills training in negotiation, interfaith dialogue, and in linking together antagonist groups in common aid projects. But when it comes to anything that seemingly relates to the military, these organizations tend to get nervous.

This is a subset of the gap between the role of the military and aid organizations in complex emergencies. International NGOs and UN agencies struggle to make their role purely humanitarian. They do not want to "muddy" their image with a local population by being too closely associated with the military. To do so, they feel, compromises their neutrality (to the extent that they are perceived as being neutral) and endangers their staff members.

I can relate to this desire to keep a clear demarcation between those doing humanitarian relief and the military. After major fighting of the war in Iraq ended in 2003, American Refugee Committee decided to enter the Maysan Governorate (in the south) to establish a humanitarian relief program with a headquarters in Amara. Our information indicated that the southern part of Iraq was relatively stable. We asked our director of security to go to Kuwait City and then travel by land into Amara to determine whether a program could be established. He was accompanied by a team of emergency aid workers, whom we had recruited soon after the start of the war.

We stayed in contact with our staff members on a daily basis using a satellite phone and email, made possible through a communications center set up by the British military. The British were the occupiers of the Maysan Governorate.

We began to hear conflicting stories from our staff members. One would say that there was adequate security in which to operate. Another would say there was inadequate security, and that one of our staff members was carrying a gun. As a result, those of us at headquarters decided that I needed to go to Amara to assess the situation.

I flew into Kuwait City, got a ride to the border of Iraq, and met one of our staff members at the border. My first indication that the security environment was problematic was when I learned that the trunk of the car in which we were riding was filled with machine guns. The second sign was the unusually fast speed at which we were traveling, which, I was told, was the way to evade an attack by gunfire or an improvised explosive device (IED).

After spending about a week in Amara, we learned that the British military was interested in having our organization provide management and oversight support for the rehabilitation of a water system in a nearby village. The British had lost three military police in that village a few days earlier. A mob had formed around the police station demanding services, most of which they had lost during the U.S.-led invasion. One of the military police officers mishandled the crowd by striking (not killing) those who became aggressive. All three of the officers were then attacked by the mob, tortured, and shot in the back of their heads. Rather than being vindictive, the British military decided to quickly show tangible signs of progress in reinstating services. Hence, it wanted to rehabilitate the village water system.

We decided that we would go out to the village to see how we could assist. I asked some of our local staff members, mainly engineers, to come with us. We drove to the British military compound in our rather unassuming small pickup truck. We learned that the British had a standing order that any soldier going into that village needed to go with substantial armored protection. Soon thereafter, a British tank rolled into the area where we were, and we learned that the tank was going to lead our group into the village.

It was surreal, as a humanitarian worker, to be riding in a truck following a tank into a village where soldiers had been tortured and killed days before. (Looking back on it, I should have refused to go, but I was so surprised by the tank that I, regrettably, said nothing.) We stopped a number of times to look over the water plant and other areas where pipes had been damaged. And, in each case, soldiers with machine guns and flak jackets surrounded us.

In my work, this is the most acute example of failing to maintain the gap between the military and humanitarian aid workers. Needless to say, we shut down the program.

If international humanitarian organizations have staff members wearing camouflaged fatigues, looking essentially like special forces teams without the guns, the local population and controlling powers will likely view them as linked to a military unit. That is one reason Barrs argues that preparedness support and mobile aid are probably best provided by specialized NGOs. Such division of labor can prevent blurring the line between humanitarian workers and military cadres.

Other Challenges

Preparedness support must have the confidence of the local populace and the local controlling powers (government officials, warlords, or rebel groups, among others). This requires acceptance or at least tolerant indifference of local controlling powers. One way to implement such a program is to start early before violence—prior to when violent groups might object to people being trained in preparedness support or Mobile Aid, or both.[11]

But make no mistake, highly insecure environments often lack operating communication infrastructure, making early warning difficult even when local populations are prepared to engage in mitigation or to flee. In some cases, communication systems must be set up anew. An example is the one being developed in Burma by the Genocide Intervention Network (GI-NET, which is now called United to End Genocide). As the network describes it:

GI-NET is working with a local implementing partner, the Free Burma Rangers, to create a radio-based early warning system. This network enables civilians to receive and send warning information and distress calls, greatly enhancing the time they have to prepare and flee. GI-NET has begun funding a pilot program that will deliver over 200 radios and associated equipment to villages in targeted areas of Burma's Karen State. Villagers are being trained on radio use and the appropriate network protocol.

This system means that villages in the network will have very little chance of getting surprised by an attack. An early version of the radio system has already been used and saved lives. Additionally, it will not only give villages advance warning of military attacks but will also facilitate the deployment of "Relief Teams" led by the Free Burma Rangers.[12]

When moving safely "out of harm's way," it is important to decide the optimal size of a group, the best routes, locations at which to reunite, whether to travel at night or on roads, the need for and method of creating decoy tracks to throw pursuers off course, and identifying places of potential ambush. This is an area where learning from mistakes is unacceptable. The "learn as you go" approach can easily result in loss of life as people "overcome a steep learning curve" (Barrs 2010, 17–18).

Creating safe sites to which to flee can require considerable time and ingenuity. For instance, some vulnerable groups dig rooms underground, using camouflaged doors and concealed ventilation. Observation and listening posts are established so that people can detect incoming belligerents.

It is critical that people practice ahead of time what to do if they are threatened. Unless there has been training and practice beforehand, people will tend to panic, make noise, clumsily grab precious items, and flee along routes easily followed by oncoming attackers. So observed Ishmael Beah, who was captured and forced to become a child soldier. Within seconds of an attack people "started screaming and running in different directions, pushing and trampling whoever had fallen. No one had the time to take anything with them. Everyone just ran to save his or her life. . . . Families were separated [and they] left behind everything they had worked for their whole lives. . . . We had yet to learn . . . survival tactics."[13]

Summary

Early warning and early response approaches often have treated the local population as if they are passive recipients of informed, brave, and well-prepared interventions from the outside. It is preferable to have a "people-centered" approach. Two ways of supporting vulnerable populations is to prepare them to be able to flee or hibernate, on the one hand, and to train them to provide aid to members of their communities when living in insecure environments, on the other.

It is sobering to consider that international humanitarian organizations can inadvertently create a false sense of security among vulnerable people. It is therefore necessary to develop approaches that will reduce that vulnerability and to design early warning systems that not only identify when an attack is likely, but also assess the extent to which attackers can reasonably be expected to be convinced not to attack.

Because fleeing danger is disruptive and traumatic, people need to be trained how to lead a timely evacuation. This leadership can be made more effective by helping to create a higher degree of comfort in fleeing through evacuation drills, accompanied with cultivating an understanding about how people can stay alive after fleeing. People must be able to recognize the danger with which they are faced while envisioning survival in a new place.

There are multiple responses possible to an early warning of potential violence. Certainly, intervening to prevent bloodshed is one. But we would be remiss to ignore that there are numerous circumstances when evacuating is an optimal choice in saving human lives.

III Resource Allocation Considerations and Recommendations

The rapid spread and visible success of information technology, or more accurately the success of two particular technologies, the Internet and mobile phones, has helped to construct a utopian vision of how technology will change the world. More than a few people . . . have assumed that these changes will necessarily lead to greater efficiency and greater power. We're increasingly aware that this is not the case, that there are limits to what technology can achieve.

—Paul Currion, "Conclusion"

This part of the book covers various criticisms of and recommendations for conflict early warning and early response systems. The criticisms, covered in chapter 11, fall into three categories: (1) they do not focus on the causes of the violence; (2) they are a drain on resources (both human and financial), to the detriment of other programs that are of higher priority; and (3) there is often a lack of a mandate to intervene.

Chapter 12 covers future directions and recommendations, arguing that there are ways to keep costs down, that it is important to exploit economies of scale. It describes how decision sciences can assist the field of conflict early warning and early response in refining our technology of when to issue a warning. It also provides guidance on how to minimize costs. Finally, it offers recommendations for the field of violence prevention for funders, researchers, and practitioners.

While writing this book, I have had numerous people ask me my opinion of using technology for violence prevention. I provide them with a brief assessment of the limits—but also the merits—of using technologies, depending upon the circumstances in which the technologies are applied. I find that those conversant with technologies (especially social media) are optimistic and inquisitive about different innovative approaches. Conversely, those who pride themselves in the use of email, but know little else about other ways of communicating electronically, resonate with an assessment that technologies are of limited use, that one must be careful not to get swindled by "technology snake oil salespeople." It is as if those who are behind in understanding and using new technologies feel a satisfaction in "not missing out on something important" that soothes their sense of relative technological backwardness.

With these technological and largely generational dispositions in mind, I encourage you to release yourself from emotional dispositions that distort sober assessment and healthy contemplation of potentialities. The goal of using technology for early warning and early response is to increase effectiveness of preventing violence, to reduce the vulnerability of vulnerable (and often impoverished) people, to inform decisions and enhance capacities for timely and prudent actions in situations filled with risk and uncertainty, and to do so without draining precious resources from services that are sorely needed. To use the emphatic instruction of my former high school football coach after I reliably proved incapable of catching a perfectly thrown pass: "Keep your eye on the ball!"

11 Concerns about Misallocation of Resources

One phrase from the aid world that captures [the debate over resources] is that of "the well fed dead." It was first used in Bosnia I believe when the international community was doing all it could in terms of day-to-day food/life support but did not provide corresponding protection.

—Casey Barrs, email correspondence, June 29, 2009

In some ways it is perhaps more important for us to properly measure the impact of technology, because we have more to prove. Organisations working on ICT4Peace issues are generally resource-constrained and—in brutally simplistic terms—every cent spent on technology is a cent that doesn't go to affected communities.

—Paul Currion, "Conclusion"

This chapter covers three categories of criticisms regarding resource allocation: (1) that early warning and early response programming has the wrong focus; (2) that it drains limited resources; and (3) that it lacks a mandate.

The Wrong Focus

Some argue that conflict early warning and early response fails to pay adequate attention to the *structural* causes of violence. The *operational* focus of preventing violence is seen as a distraction. According to the Carnegie Commission on Preventing Deadly Conflict, *structural prevention* measures are designed to "ensure that crises do not arise in the first place or, if they do, that they do not recur." This is distinguished from *operational prevention* that involves "measures applicable in the face of immediate crises" (1997, as quoted by Nyheim 2008, 20).

As explained earlier, violence prevention at a local level is strategic in both the early and late phases of the conflict process. In the pre-massive violence phase, it can prevent an unraveling of a conflict into massive bloodshed. That is, it can prevent the mischievous influences of *trouble-makers* or others seeking to sabotage efforts to counter conflict escalation. Similarly, in post-conflict situations during which a delicate ceasefire prevails, local conflict early warning and early response is strategic in preempting violence facilitated by *spoilers*.

But those who critique conflict early warning and early response stress the importance of asking why people are becoming violent. They are concerned that designing programs to respond to incipient violence distracts people from focusing on the changes that need to be made in the structural conditions that are creating or maintaining the conflict overall. This line of reasoning is similar to the challenge people face in addressing poverty with charity as compared to working for just local, national, and international policies to address the root causes.

There is no simple answer to this dilemma. On one hand, structural changes are needed to address the fundamental reasons behind substantial societal tension. On the other, operational violence prevention can stifle troublemakers and spoilers so that tension does not escalate. Using a stock-and-flow analysis (an approach used in system dynamics computer simulation that analyzes accumulations and rates of change; see Bock 2001a, 119–138), think of a bathtub: building intergroup goodwill based in part on structural changes is like a slow flowing, or dripping, faucet (a slow in-flow). Goodwill accumulates over time, like water does in a tub (an accumulation, a stock). Violence cultivated by troublemakers or spoilers is like opening a large drain. The goodwill that was carefully cultivated dissipates rapidly (a fast outflow). So it is defendable to provide immediate responses while also working on the systems that are at the root of the problem in the first place. Viewed this way, *structural* and *tactical* efforts can go hand in hand.

A Drain on Precious Resources

Related closely to the criticism that early warning and early response systems divert attention away from structural injustice is the argument that these systems divert resources away from other programs that are more important. In Sri Lanka, for instance, I regularly heard people in war-torn areas who suffered extreme poverty express their frustration in hearing about the need to create peace. They explained that their primary concern was in getting adequate food and shelter for their families. Similarly, data from International Rescue Committee on the civil war in the Democratic Republic of Congo indicate that for every violent death of combatants, sixty-eight died nonviolently, many from disease and malnutrition (Lacey 2005, as cited by Leaning 2010, 216). It is hard to argue with a view that building peace should not take precedence over getting potable water, food, and shelter.

Of course, this is not unlike questions and choices facing many working in humanitarian relief and development. In 1993, I visited a remote area of Somalia, just outside Baidoa. This was during a horrific famine. I was there with Catholic Relief Services (CRS). After flying from Nairobi, Kenya, in a small plane, a group of us were driven to this remote area in Toyota four-wheel-drive pickup trucks. Armed guards, who looked to be of high school age, wielded automatic weapons as they chewed "khat," leaves that serve as a stimulant (as if these young men needed any!).

We were at a food distribution site with UN peacekeepers when two people, with bloody wounds that were not healing, approached me. I could not speak their language, but I did understand that they were asking for help, pointing to their sores and pleading with me through their eyes. I asked my friend, who was running the relief program at that location, how we could help the injured woman and boy. I will never forget the expression on my friend's face when he told me the program was focusing on public health (providing potable water, food, and hygiene packs) rather than curative care. I learned that the two needed to have their limbs amputated, because their sores were infected severely, but that there was no clinic nearby. When I asked my friend what would happen to the woman and the boy, he told me they would probably die. He had a terrified look on his face. It was then that I understood what I was told back in Baltimore at CRS headquarters: sometimes in humanitarian relief work it is important to "put your heart in your back pocket." My friend, through his decision to focus on water, food, and hygiene, was saving more lives than if the focus was on clinics that could save individuals with severe ailments.

Those working and living in conflict zones, with finite money and staff members at their disposal, must make similar choices about where to direct scarce program resources. Certainly, on the one hand, saturating an impoverished area with programs aimed at preventing violent conflict when people do not have enough potable water, food, or shelter is a misallocation of resources. On the other hand, if the violence is causing their displacement and deprivation, it needs to be prevented. NGOs and CBOs clearly need to be careful not to be swept into a "humanitarian trend" or "fad" (or, as some of my coworkers call it, the "program flavor of the month"). They must collaborate with one another so that basic human needs are met, violence is mitigated, and solutions are sought to address the reasons for the deprivation and violence in the first place.

Sometimes satisfying basic needs is more important than everything else. In Sri Lanka, FCE was criticized by donors for "mission creep" when it initiated humanitarian relief following the tsunami in 2004. Those funding the early warning and early response program were concerned that this expanded programming would dilute FCE's focus on the central reason the organization had been created, its core competency: violence prevention. But knowing what I heard in the field when people expressed concern about not having food and shelter, the focus on humanitarian relief was understandable. In some respects, not helping people overcome the devastation of the tsunami would have rendered FCE's other work illegitimate in the eyes of those in extreme need.

There also is a question of the best allocation of resources to build early warning, as compared to early response, capacity. There is a sense in which there has been a misplaced emphasis on early warning at the cost of early response. Bond and Meier argue, "The predominant practice in Africa has been to invest more financial resources towards developing elaborate warning databases at the expense of building human expertise on conflict monitoring and analysis, and peacebuilding skills and techniques" (2006).

It is likely in the future we will see more extensive use of early warning and early response systems at a local level. Not only have they proven to be effective in saving lives, but also there is a huge research and development apparatus being built for data generation and pattern recognition that no doubt will spur the technical sophistication of combining events data, maps, and other disparate information for useful purposes. As explained earlier in this book, infectious disease control, and natural disaster prediction (such as famine forecasting) are but two juggernauts in this scientific stream.

Keep in mind that conflict early warning and early response has a tarnished reputation because of misapplication or poor execution of macrosystems. Nonetheless, the arguments for and against macrosystems are instructive as they relate to approaches to prevent violence locally.

Governmental development agencies have, by their actions, shown that they do not believe the benefits of early warning systems are substantial enough to warrant their expense. In a recent survey, staff members of some of these agencies stated clearly that early warning was not of particular help to them. The exception is qualitative early warning systems (such as the one being implemented by International Crisis Committee). This, according to Nyheim, is probably due to the value of this information during development agencies' planning cycles. Whereas the fact that Swiss Development Cooperation discontinued funding FAST is indicative that development agencies are likely to find quantitative early warning systems of limited value (2008, 25, 32).

But Yiu and Mabey reached a strikingly different conclusion, even when referring to risk assessment (not early warning) at a macrolevel. They wrote:

Preliminary analysis suggests that conflict prevention is significantly cheaper than reacting to crises once they have emerged. Therefore, the failure of a risk assessment process to predict a crisis (false negative) is more costly to decision makers than over-prediction of crisis (false positive). Analysis suggests that even a 25 percent successful identification rate for emergence of violent conflict would be accurate enough to make significant preventative action cost-effective. (2005, 1)

Even more striking, swisspeace and the Alliance for Peacebuilding claim that the average cost of conflict prevention is less than 2 percent of the cost of war. These figures reflect a broad view of governmental and multi-lateral budgets, not solely those of development agencies and aid ministries. Similarly, Nyheim found in his study that systems on the local level were helpful to the U.S. government's office of the Coordinator for Reconstruction and Stabilization, and the French government's Secretariat General de la Defense's Systeme d'Alerte Preconce (2008).

A contrary view is that making hypothetical assumptions about effectiveness leads to spurious conclusions. As Austin, Brusset, Chalmers, and Pierce write:

Existing studies on the cost-effectiveness of prevention tend to point to cases where *ex post facto* the cost of inaction considerably outweighed the cost of hypothetical conflict prevention. However, the results of such analysis can often be misleading, as the results of an action can never be known in advance, and so the *estimated* costs are likely to be very different from the *actual* costs. Conversely, the benefits of prevention are also unknown, preventable actions may be unsuccessful or may simply delay the onset of violence. Alternatively, even without preventive action, conflict may not have taken place. While the general argument for more resources for prevention and peace building has considerable attraction at first sight, therefore, a more rigorous approach to estimating its costs and benefits is needed in order to be able to operationalize the concept. (Austin et al. 2004, 68n106)

There are, of course, ways in which the costs of local systems can be brought down considerably. As covered earlier, one way is to automate data generation from media sources. Yiu and Mabey provide a helpful explanation in their report to the Prime Minister's Strategy Unit of the British government:

New IT systems make possible automated tracking and coding of large quantities of news sources into indicators of instability (e.g., increased incidence of reports of communal violence or hate speech in local newspapers). This is a relatively cheap way of combining high volumes of data into "real time" consistent and quantifiable indicators which can be compared over time to see escalating risks, supplementing reporting from UK [diplomatic] Posts. Systems have already been developed by NGOs and the private sector, and are being used currently by the European Commission. The US Army has also recently developed a sophisticated system called [Forecasting of Crises and Instability using Text-based Events] FORECITE.[1] (2005, 33)

Another way to bring down costs, as we noted in chapter 6, is through crowdsourcing. Following the earthquake in Haiti on January 12, 2010, Ushahidi worked with a team of students from both Tufts University and Stanford University to build an Internet system for processing text messages received from Haiti at a rate of one thousand to two thousand a day. Over one thousand volunteers from the United States and Canada, many of whom were part of the Haitian diaspora, translated and geocoded this information, posting it on a digital map after receiving a report, within an average of ten minutes. The cost was virtually nil, with the exception of $97,000 paid to Ushahidi's university-based teams, which team members used, in part, to travel to Haiti to get feedback on how to enhance the usefulness of digital maps and reports (Heinzelman and Waters 2010, 8).

But we should not lose sight of the value of supporting local capacity in our zealotry for more useful and less expensive technologies for early warning. Arguably, the most cost-effective component of early warning and early response systems at a local level is in cultivating and assisting CBOs. This is easier said than done. Field staff members must listen to those within a community with whom they have rapport and, in a sense, "go with the flow." Supporting local capacity is not a linear, straightforward, process. This excerpt, among others, of an impact assessment report of FCE's Human Security Program illustrates this point well:

While the robustness of FCE's Co-existence Committees varies from one location to another, it is clear that some do exist and they constitute multi-ethnic networks that are having an impact [on] . . . violence prevention. . . . FCE's inter-ethnic staff and formalized Sinhala-Muslim-Tamil networks of "connectors" in the districts are one of its most effective peace building assets.[2] The FCE Co-existence Committees are uneven in terms of the kinds of committees that exist in each district, in their effectiveness, and in the frequency of their meetings and interventions. This reflects variance in the terrain of political violence in each district. Some only meet when necessary (i.e. the Co-existence Committee of Cattle Owners in Ampara only meets when a conflict arises or when decisions need to be made about grazing cattle).

Others attempt to meet on a monthly basis (such as the Co-existence Committee of Community Leaders in Batticaloa), but they have problems with attendance when violence and fear increases. The Co-existence Committee of Religious Leaders in Ampara and Batticaloa meets when there is a need to calm a violent event. Their members come from Mosque Federations, and Buddhist leaders and Catholic priests have been involved as well. There was a peace event in Batticaloa in December when FCE members visited a church, mosque, kovil [a Tamil name for a Hindu temple], and pansala [living quarters of a Buddhist monk] in a procession that was organized by Catholic, Muslim, Hindu and Buddhist religious leaders and FCE. The Women's Co-existence Committees are quite different from one district to the next. In Ampara the committee is both new and has only a handful of members; in Batticaloa it has 90 members and is very active/assertive. (Bock, Lawrence, and Gaasbeek 2009, 214)

A common and arguably valid criticism of early warning and early response systems is that they place too much emphasis on outsiders and on technology to the detriment of supporting local capacity. It is important to keep this criticism in mind while at the same time identifying what can be done by outsiders and by technology that is helpful. What can be built on a foundation of people at a local level who look for warning signs and are prepared to intervene to prevent violence or to physically evade it?

Lack of a Mandate

Another critique leveled against early warning and early response proponents is that vulnerable people cannot trust that appeals to mid- and top-level leaders will result in interventions to prevent violence. Peacekeepers, even if present, often lack a mandate to prevent violence. This question about a mandate to intervene has been raised with numerous transnational organizations, such as the Organization of Security and Co-operation in Europe (OSCE), and with UN peacekeeping forces.

Bureaucratic clumsiness can result from the right hand not knowing what the left hand is doing. It is very difficult for commanders of various military bodies to coordinate their troops in *complex emergencies* (those involving both humanitarian crises and violent conflict). And an attack on a small village by a renegade splinter group, especially along a region that borders different areas of responsibility ("AORs" as the military calls them) can fall between the cracks.

Early warning and early response is rightly criticized when peacekeepers are present, when there are obvious warnings that innocent people are likely to be attacked, and when they are attacked and the peacekeepers do nothing. This happened during the genocide in Rwanda in 1994. Those fleeing violence moved to an area near UN peacekeepers. The peacekeepers, lacking a clear mandate, were then ordered to leave. The fleeing Rwandans, mainly Tutsis but also some Hutus who tried to hide Tutsis, were subsequently slaughtered. According to the British Broadcasting Corporation, the commander of the UN peacekeeping force, Lieutenant General Romeo Dallaire, told a conference at the UN's headquarters in New York commemorating the tragedy that that "no one was interested in saving Rwandans." He commented sadly that there continued to be a lack of political will to give UN peacekeepers a mandate to intervene. He stated: "I still believe that if an organisation decided to wipe out the 320 mountain guerrillas there would be still more of a reaction by the international community to curtail or to stop that than there would be still today in attempting to protect thousands of human beings being slaughtered in the same country."[3]

But there are instances when peacekeepers have deliberately gone beyond mandates. An example is the Sri Lankan Monitoring Mission (SLMM). It was formed in 2002 to monitor compliance to the ceasefire agreement between the government of Sri Lanka and the Tamil Tigers. Norway, Sweden, Finland, Denmark, and Iceland deployed monitors throughout the country. Their mandate was strictly to monitor and then to report on violations of the agreement. But, when I visited the Eastern Province of Sri Lanka, I found that they often deliberately exceeded their mandate. I learned that in a number of instances when they heard about violent conflict that was about to occur they drove their four-wheel-drive pickups trucks, brandishing huge flags with the initials "SLMM" on them, into potential battle zones on the theory that an international presence could potentially nip violence in the bud. And, more important, I learned that when they reached conflict zones before violence occurred, they were generally successful; their mere presence was an effective deterrent. They mediated disputes and spoke with extremist leaders, successfully persuading them to not attack. When I interviewed a number of the SLMM team leaders, they stated with some trepidation that they knew they were exceeding their mandate. But I had the sense that they had unofficial permission to engage in such activities.

Summary

There are three categories of criticisms leveled at early warning and early response systems covered in this chapter: that the focus of programming should be on structural causes of violence; the systems are a drain on resources; and there is often no mandate to intervene.

It is important to keep in mind, however, that violence prevention at a local level is strategic in both the early and late phases of the conflict process. In the pre-massive-violence phase, violence prevention can avert an unraveling of a conflict into massive bloodshed. That is, violence prevention can counteract the mischievous influences of troublemakers or others seeking to sabotage efforts to counter conflict escalation. Similarly, in post-conflict situations during which a delicate ceasefire prevails, local conflict early warning and early response is strategic in preempting violence facilitated by spoilers. But those who critique conflict early warning and early response are concerned that designing programs to respond to incipient violence distracts people from focusing on the changes that need to be made in the structural conditions that are creating or maintaining the conflict overall. There is no simple answer to this dilemma. On the one hand, structural changes are needed to address the fundamental reasons behind substantial societal tension. On the other, operational violence prevention can stifle troublemakers and spoilers so that tension does not escalate.

Another argument is that these systems divert resources away from programs that are more important. It is hard to argue with a view that building peace should take precedence over helping people to get potable water, food, and shelter. Those working and living in conflict zones, with finite money and staff members at their disposal, must make difficult choices about where to use scarce program resources. Certainly, saturating an impoverished area with programs aimed at preventing violent conflict when people's basic needs are not met is a misallocation of resources. Yet if the violence is causing their displacement and deprivation, it needs to be addressed.

Sometimes satisfying basic needs is more important than everything else. Not helping people overcome the devastation of a disaster renders other work illegitimate in the eyes of those in extreme need. There are, of course, ways in which the costs of local early warning and early response systems can be brought down considerably. One way is to automate data generation of media sources. Another is through crowdsourcing. But we should not lose sight of the value of supporting local capacity in our zealotry for more useful and less expensive technologies for early warning. Arguably, the most cost-effective component of early warning and early response systems at a local level is in cultivating and supporting CBOs. This is easier said than done. Field staff members must listen to those within a community with whom they have rapport, and, in a sense, "go with the flow." Supporting local capacity is not a linear, straightforward, process.

A common and arguably valid criticism of early warning and early response systems is that they place too much emphasis on outsiders and on technology to the detriment of supporting local capacity. It is important to keep this criticism in mind while at the same time identifying what can be done by outsiders and by technology that can support local capacity.

Another critique leveled against early warning and early response proponents is that vulnerable people cannot trust that appeals to mid- and top-level leaders will result in interventions to prevent violence. Peacekeepers, even if present, often lack a mandate to prevent violence. Bureaucratic clumsiness can result from the right hand not knowing what the left hand is doing. It is very difficult for commanders of various military bodies to coordinate their troops in *complex emergencies* (those involving both humanitarian crises and violent conflict). And an attack on a small village by a renegade splinter group, especially along a region that borders different areas of responsibility ("AORs" as the military calls them) can fall between the cracks.

Early warning and early response is rightly criticized when peacekeepers are present, when there are obvious warnings that innocent people are likely to be attacked, and when they are attacked and the peacekeepers do nothing. This happened during the genocide in Rwanda in 1994.

But there are instances when peacekeepers have deliberately gone beyond mandates. An example is the Sri Lankan Monitoring Mission (SLMM, formed in 2002 to monitor compliance to the ceasefire agreement between the government of Sri Lanka and the Tamil Tigers. Norway, Sweden, Finland, Denmark, and Iceland deployed monitors throughout the country. Their mandate was strictly to monitor and then to report on violations of the agreement. But, when I visited the Eastern Province of Sri Lanka, I found that they often deliberately exceeded their mandate.

12 Future Directions and Recommendations

This chapter offers ideas about how to enhance software as it relates to the volume of data that will be available for early warning. It describes how decision sciences can assist the field in refining our technology of when to issue a warning. It provides guidance on how to minimize costs, and offers recommendations for the field of violence prevention for funders, researchers, and practitioners.

Exploiting Economies of Scale

While circumstances require that early warning and early response systems are tailor-made from one location to another, there are components that can be developed and used globally. They fall into these categories: software, training programs, and networks.

To the extent that we use computers to track events, efficient and relatively accurate software will be required. For example, it would be helpful for an application ("app") to be developed for cell phones and notebooks that allow NGOs and CBOs to interrogate data, with functionality similar to that used by the Philadelphia Police Department with Hunchlab.

Training programs, as well, can be developed with an eye toward serving as a basic foundation globally while being adapted for local circumstances. For instance, computer simulation training games on intervention and evacuation techniques can be designed for use over the Internet in multiple languages. Venues for training can be created using virtual reality sites in which scenes can be created and event-driven simulations conducted using avatars. As we have seen, CeaseFire commissioned a Second Life site for training people in violence prevention.[1]

The more difficult technological challenge lies in the lack of cell phone towers in remote locations where there is considerable violence. Of course, substantial strides have been made with cell phone networks. Where there are no cell phone towers it is, of course, possible to use satellite phones and radios. Satellite phones currently are prohibitively expensive for many NGOs, and certainly for community-based organizations. There are promising efforts to reduce the cost of satellite connections for those exchanging data as compared to voice and video. One also wonders about expanding data transmission using "lower-tech" approaches such as ham radios, localized radio stations, and the like. We have seen how CEWARN is developing such radio technology. Inventors will no doubt develop completely new forms of data transmission that are relatively inexpensive, with the price going down exponentially the more the technology is used.

Even in locations replete with cell phone service and high speed Internet, governmental interference during volatile periods can limit connectivity. Following elections in Iran in 2009, the government blocked many forms of electronic communication in an effort to stifle protests that the election results had been manipulated. Interestingly, protestors were able to continue communicating with each other and the outside world using various Internet-based communications platforms like Twitter, though they too were subject to censorship.[2]

Governments' control over communications in undemocratic societies undercuts the view that cell phones and the Internet are "technologies of liberation." Morozov, in his provocative book *The Net Delusion*, argues that oppressive governments are able to jam electronic communications with great effectiveness, describing the "dark side of internet freedom" (2011). Such censorship is common in dictatorships but, as Howard notes insightfully: "Censors often seem one step behind and reactive, developing restrictions in response to creative maneuvering by citizens armed with mobile-phone cameras, portable flash drives, and basic knowledge of how to use free internet tools" (2011, 157).

Enhancing Methods and Software for Generating and Securing Data

The use of text messaging is a breakthrough for local conflict early warning and early response. There are other approaches to generating data, however, that are on the horizon. Here, seven are covered that involve (1) categorizing and queuing; (2) expanded use of natural language interface; (3) verification of statements by key leaders; (4) translation software; (5) unstructured data feeds; (6) image recognition; and (7) using the stars.

The first approach, *categorizing and queuing*, was developed by the European Commission's Joint Research Center (JRC) and has been adapted for the African Union's Continental Early Warning System (CEWS). A European

Media Monitor (EMM) program has been converted into an Africa Media Monitor (AMM). It searches for key words using sophisticated mathematics not unlike those employed by online search engines (such as Yahoo and Google). Once the media monitor identifies key words in an article, it categorizes it. Analysts then read these articles. In a sense, then, this approach creates a quasi-assembly line. It quickens the process, and it works in any language. Keep in mind, however, that it requires significant human resources.[3]

A second approach, *expanded use of natural language interface*, involves "answering machine" software which can be integrated into phone systems so that people can report incidents by phone, choosing a category of an event from a menu.[4] They can also choose to whom to send a report, by selecting from preestablished lists of recipients (of those preapproved to receive warnings). Of course, such phone systems can be password protected. FCE developed a partial version of this approach. Freedom Fone in Zimbabwe is also developing this functionality.[5]

A relatively rudimentary capability of such answering systems involves short statements being coded. For instance, if a field officer is calling in an incident report while driving (not to encourage this dangerous practice), he or she could say the name of the event category rather than press a specific number on the phone to identify it. He or she could say "abduction," for instance, rather than pressing a preassigned number.

As natural language processing software becomes more accurate, it is increasingly possible to generate data from speech. This might be especially useful when monitoring religious incitement. While there is a disquieting "big brother" surveillance flavor to this if a government does it (in the Orwellian sense of the term),[6] microphones could be installed in places of worship. Natural language interface software could convert sermons and other public statements into sentences. And software like the Virtual Research Associates (VRA) Reader could be adapted to convert those sentences into events data.

Interestingly, a third approach is being developed, again by stock traders (from whom we have learned the value of analyzing both fundamental and technical indicators), which involves *verification of statements by key leaders*. Their verbal communications can be analyzed using computerized lie detectors (Bowley 2010, B4). While this might appear to be an overly futuristic, impractical, and expensive technology, especially for violence prevention at the local level, one can imagine how such an approach would be helpful when monitoring hate speech and other forms of incitement, in evaluating whether a threat is a bluff, and in determining if a rumor is based on fact.

A fourth approach involves *language translation*. As reported earlier, automated events data generation currently is limited to the English language. It is conceivable that resources will be brought to bear to translate the software into other major languages such as French, Spanish, Arabic, and Hindi. Keep in mind that in many conflict zones throughout the world, other languages and dialects are used, and it is from communication in those dialects that some of the most useful information can be gleaned. The economies of scale, however, are unfavorable for these dialects, rendering translation too expensive.

Financing of language conversion, as with the use of lie-detector software, probably will target potentially profitable business applications (Nyheim 2008, 9). Because language conversion into regional dialects is not likely to be of interest to companies, this is an area that will need support from foundations, governments, and transnational organizations.

Translating the VRA Reader into other languages is likely to be more expensive than translating text from, say, French to English. Commercial automated translation software is available for major languages. Converting text from one language into English and then using the VRA Reader to generate events data from it is probably less accurate than the original text in English, but such conversions would be relatively inexpensive since translation software is already available.[7]

The need for language conversion software, however, would be reduced significantly through the fifth approach, sometimes referred to as *unstructured data feeds* (Bowley 2010, B4). Crimson Hexagon and other companies are developing this technique using both parsing and "stemming" on the Internet.[8] Crimson Hexagon software is "taught" what to look for by scanning from ten to thirty blog entries. These entries are run through a parser that uses stemming, which Meier explains in his account of monitoring bloggers and their opinions on riots in Tehran:

Every word in a given text is reduced to its root word. For example, "rioting," "riots," "rioters," etc., [each] is reduced to riot. The technology creates a vector of still more just to characterize each blog entry so that thousands of Iranian blogs can be automatically compared. By providing the algorithm with a sample of ten or more blogs on, say, positive perceptions of rioting in Tehran were this happening now, the technology would be able to quantify the liberal Iranian bloggers' changing opinion on the rioting in real time by aggregating custom vectors. . . . Instead of trying to find the needle in the haystack as it were, the technology seeks to characterize the haystack with astonishing reliability such that any changes in the haystack (amount of hay, density, structure) can be immediately picked up by the parser in real time.

Furthermore, the technology can parse any language, say Farsi, just as long as the sample blogs provided are in Farsi.[9]

A sixth approach involves *image recognition*, providing at least a part of a solution to the challenge of obtaining accurate location data. Image analysis is already being developed by Marya Lieberman and a team of undergraduates at University of Notre Dame, along with Jennifer Chen of Northwestern University (who is also part of Ushahidi and an active participant in the Harvard Humanitarian Initiative). Lieberman and her students are developing a cell phone application with which to analyze whether pharmaceuticals are counterfeit or impotent, or both. Lieberman impregnates paper with chemicals. A capsule of a drug, such as diethylcarbamazine citrate (DEC), which is used to prevent lymphatic filariasis (commonly known as "elephantiasis"), is broken in half and then rubbed against treated paper. The paper is then dipped into water and shades of colors appear. The customer then takes a picture of the shades of color using a cell phone, while at the same time explaining in a text message the source of the drug (such as a pharmacy or street vender) and the location of purchase. The text and image are then sent to a central location. A central computer analyzes the image to determine purity. The information is then plotted on a digital map that depicts drug purity, or lack thereof, in a given region.[10]

How might this approach be used for violence prevention? It is conceivable that images of events and of major landmarks can be used to create events data, on the one hand, and the approximate location of the event, on the other. A text message could be used instead, of course, for both, but a picture of an event is likely to convey the reality of the event, and a picture of a landmark is a quick way of conveying location. Just as people are being asked to enter codes for event types, they also can be asked to take a photo of a major landmark. Ideally, image analysis can be automated for approximate location data. Similarly, for largely illiterate populations such as those in Afghanistan, symbols of categories of events could be distributed on paper. When people witness an event, they can then take a picture of the image and send it to a central computer that analyzes the image and, if location data are available, plots the event onto a map. Such an approach, when combined with GIS data, can be used in land disputes, solving partially the problem faced by those implementing alternative dispute resolution programs. The nonprofit Internet Bar Organization (IBO) is facing this problem in its Silk Road Initiative. As Himelfarb and Paradi-Guilford explain:

Land ownership is a major source of conflict in Afghanistan, particularly in remote and rural regions. Eighty-five percent of Afghans rely on the land's resources as their primary source of livelihood. . . . The Internet Bar Organization (IBO), a nonprofit organization that focuses on the use of technology tools for justice and dispute resolution, is developing the Silk Road Initiative, a project that will train Afghan dispute specialists to leverage mobile telephony to collect information vital to resolving land disputes through peaceful arbitration. This includes using smart phones to record GPS coordinates of boundaries, taking photographs, noting information about the disputed properties, and then sending the data via SMS to Silk Road's main hub for arbitration. This information will be kept in a digital repository with free and open access. In addition, a panel of arbitrators (either trained by IBO or drawn from local elders) will be empowered to render decisions, which will be transmitted back to disputants via mobile phones. (2010, 10)

And, finally, using a sextant to *determine location* has been adapted for cell phones, similar to how ships were circumnavigated in bygone years. The app provides rough coordinates based on the time of day and the location of the sun, or, at night, the moon and stars. One such application, called Spyglass, is capable of approximating distance to objects. This technology, if developed further, can potentially provide relatively accurate location data with devices that do not determine GIS coordinates.[11]

Developing Decision Support Software for Early Warning

Individuals and groups working on violence prevention at a local level might well benefit from decision support software that incorporates pattern recognition, mapping, decision protocols, and data validation capacity. The features of any decision support system to be used for determining when to issue an early warning might include, for instance, categories of warnings to issue (to evacuate, to hibernate, to be on high alert, etc.), any alternatives for each category (evacuate as a large group by night down a river, by day in a plane, as small groups on bicycles dispersed in many directions, on foot through the forest in small groups, etc.), probabilities of violence occurring, the amount of time likely to be available for a response, and the potential consequences of responding or not responding.[12]

Computer scientists have developed *expert systems* that, as suggested by the name, seek to replicate insights from "experts." As Flake defines such systems, they are a "special program that resembles a collection of 'if . . . then' rules. The rules usually represent knowledge contained by a domain expert (such as a physician adept at diagnosis) and can be used to simulate how a human expert would perform a task" (1998, 451). Such systems are based on artificial intelligence software that captures associations using fuzzy logic.

It is probably post-Enlightenment hubris to believe that an expert system can provide decision support for when to issue warnings. But it is an idea worth considering as our understanding of conflict early warning and early response at a local level is enhanced. One must recognize, however, as explained earlier, that there is no substitute for human induction.

If a decision support system can be developed that pulls disparate information together in a coherent way, then the likelihood of improvements in early warning and early response precision is enhanced. A realistic goal for further development of decision support software is to provide timely integration of disparate data that will support human inductive reasoning, with experts offering insights that relate to a similar situation.

Orchestrating Visualization and Induction

The adage that "a picture is worth a thousand words" rings true when one sees how still photographs or video are posted on the Internet, thereby garnering considerable attention for what people living amid the horrors of violence are experiencing. If data are added to those images in a visual format, often involving maps, then more of the story becomes evident. As aptly put by Jennifer Leaning:

Visualization enhances the capacity to engage with large data sets, apprehend their main findings, and derive from that visual depiction a sense of patterns. . . . Through the use of patterns, informed and experienced observers can interrogate an "apparent" fact picture, generate deeply relevant questions from that interrogation, and through further research, drive closer to an understanding of what is actually going on, in a given place and time. (2010, 192)

As was covered earlier, FCE used its events data to graph trends over time. Ushahidi plots information from text messages, social media feeds, and Internet forms onto a digital map. Both examples show how visualization can enhance inductive reasoning. But the idea that users can interrogate data, as Leaning suggests, is another analytical dimension. As people seek to make sense of what is going on, or what might happen, they develop "educated guesses" or hunches. They can add additional data, expand the scope of the map, or extend trend analyses further into the past. They can add new types of data to their visual tools.

A helpful example of bringing various forms of data into the picture comes from Philadelphia, where, as explained previously, police officers combine visualization with inductive reasoning. Azavea, a company initiated by former students at University of Pennsylvania's Cartographic Modeling Laboratory, and consultant to the Philadelphia Police Department, used a "hypergeometric" approach to create the Crime Spike Detector. Renamed "Hunchlab," this software illustrates Leaning's point about how people make educated guesses and then seek to confirm or reject them based on multiple types of data and queries (interrogations) of those data. With Hunchlab, users compare recent to past crime data. The software creates a spatiotemporal grid to produce a p-value lattice (essentially a three-dimensional circular array that represents events data over time with patterns recognized and assessed for their respective likelihood of being incorrect). A p-value, which also is called a measure of "statistical significance," is the statistical probability of getting a more extreme result than the one derived through the test.[13]

With Hunchlab, events are analyzed both spatially and temporally to "distinguish statistically significant density changes." But space and time measurements are not rigid. Instead, Hunchlab allows users to set values for space and time as parameters in their hypergeometric equation that is the heart of the system (Cheetham, Felcan, and Smith 2006, 3).

Officers using the Crime Analysis and Mapping Unit's system can generate three types of reports interactively. First, a web-based system allows officers to generate statistics, charts, and other data sets needed for their specific analytical purposes. Second, an interface with data of the Bureau of Alcohol, Tobacco, Firearms and Explosives (a U.S. federal law enforcement organization) lets officers trace firearm requests electronically. And, third, the officers can use Hunchlab to "mine" data and identify patterns.[14]

Any officer of the Philadelphia Police Department can interrogate Hunchlab, querying the system quickly and sharing the results with others, thereby enhancing the department's analytical efficiency. According to Theodore, "Instead of dozens of individuals combing through thousands of pages of documents or volumes of spreadsheets and digital forms stored in multiple locations. . . . [Hunchlab allows officers to use the department's] data stores . . . [so they] can make better decisions" (2009, 1).

What does an approach like the one used with Hunchlab portend for the future of early warning and early response to violence? We can expect more data integration, visualization, and interrogation. We must, however, be careful to keep our focus on what will be most helpful to those at a local level who have the most to lose if violence erupts.

Conducting Additional Research

More research is needed into how to support local capacity, what kinds of interventions are most effective, and how much time either insiders or outsiders can take to intervene before massive harm ensues. This book is largely exploratory in each of these critical areas.

Additional quantitative studies should be conducted. While it is not necessary to be precise as to the exact minute or hour when measuring the length of time between a precipitating event and the onset of violence in different types of conflict settings, it is important that other research be done to determine the order of magnitude that reasonably can be antici-pated. Time difference—days, weeks, or months—can limit severely the actors who can intervene to prevent bloodshed. In this book, only a preliminary analysis has been reported that suggests strongly that local actors must be part of early response mechanisms for local-level violence prevention.

While it is promising that crowdsourcing techniques facilitate participa-tion in monitoring events, we know very little about how to facilitate constructive activism to bring people from passive information collection and reporting into a rather unpleasant and sometimes dangerous activity of intervention. Much fruitful research is already being done on the extent to which there is constructive collective action by those reporting human rights abuses with text messages (see Land 2009), but more is needed related specifically to intervention to prevent violence. To what extent can theories of *collective action* (developed initially by Olson 1971) and *diffusion of responsibility* (developed by Darley and Latané 1968; Darley 2000; see Fischer et al. 2011 for a meta-analysis) be used as a guide for programming?

Olson identified the "free-rider problem" whereby collective goods can be enjoyed by those who do not help to bring them about. Almost everyone in a group contributes to improving a situation, but those who do not are still able to enjoy the collective goods. There is no penalty for their "free ride." Violence prevention, of course, is a collective good, but initiating interventions can require courage and charisma. Not everyone will be inclined to intervene, but they will benefit from the avoidance of violence. There is therefore a dispositional component that must be added to this collective goods dilemma, in a leadership-followership formulation. Moreover, as Russell Hardin (1982) explains, it is important to clarify the incentives of subgroups collectively taking action within the larger group.

The social psychology of diffusion of responsibility was developed initially when people witnessed an attacker harming another individual. As long as there were many people witnessing the violence, people tended to not intervene to help the victim, thinking that it was someone else's responsibility since there were others in their midst. People tended to be bystanders. Is this phenomenon amplified in situations where one group is attacking members of another? Is there a similar cognitive process whereby people feel no need to "stick their necks out" in a situation of high intergroup tension?

To the extent that bounded crowdsourcing, bounded crowd feeding, and restricted feeding are used, it is likely that the literature on leadership can offer insights about how trust network members can mobilize people into action. Leadership can make a pivotal difference between violent or nonviolent mobilization, as we witnessed in the prodemocracy demonstrations in Tunisia and Egypt in early 2011. To what extent does leadership training, and preparations for nonviolence, need to be combined for those in a trust network who can serve as catalysts for transformative collective action?

Establishing Violence Prevention Programs in Applicable Locations

Building a peaceful society in locations of considerable tension requires years of painstaking work. As mentioned earlier, allocating resources for conflict early warning and early response is comparable to installing a fire detection and suppression system in an expensive building or in a city. It is important to ensure that they are put into place, especially in pre- and post-conflict situations where there has been massive violence.

Here are some recommended questions for those considering establishing a local early warning and early response program:

• To what extent is your location fragile? Are you in a pre-massive violence or a post-conflict phase? If so, how ruinous would it be if sporadic violence escalates?

• Can you isolate troubled areas and facilitate the development of peace committees in those locations?

• If you already have relief and development programs in operation, how costly would it be—in terms of both human and financial resources—to develop early warning and early response capacity by integrating it with other programs?

• Are there reliable media sources in English that can be used for automated data generation? Are media in other languages important to monitor? If so, how costly would it be to hire enough people to generate and validate events data from them?

• Are cell phones used widely? Given the kind of conflict with which you are struggling, do you consider it likely that you can quickly build a "critical mass" of people who will be willing to send text messages, social media feeds, and Internet form submissions when they witness an event? How would you spread the word about the event mapping system?

• Assuming a crowdsourcing approach will be used, what programs will be needed to facilitate intervention by at least some of those involved in reporting events? Will training suffice, or will highly specialized and seasoned practitioners be needed? Will it be important to hire field monitors?

• In situations in which injustice is pervasive, what would be required for a successful strategic nonviolence initiative? Have you reached out to leaders from other locations who have had success? What strategies and tactics do they recommend? What kind of violence prevention infrastructure will you create? How will you build the movement and under what conditions would it be most effective to launch it? Who will you include in your trust network?

Funding Violence Prevention Projects More Generously

In terms of funding for building peace, there has been a much greater emphasis in building intergroup goodwill than in creating a group of dedicated people to report and respond when tensions run high. As covered in chapter 11, a system dynamics perspective is that goodwill is an accumulation (or a stock). Increasing goodwill is a rate (or a flow), as is the depletion of goodwill. But the rates are asymmetrical in their respective speeds.[15]

Humanitarian organizations—branches of the UN, INGOs, and NGOS—that are working in areas of considerable intergroup tension should consider devoting a portion of program resources to early warning and early response. A collaborative entity should be created so that organizational initiatives do not work at cross-purposes. This entity should serve as a forum for sharing information, including on anticipated violence, as a way of fostering the "wisdom of the crowds."

Typically, UN organizations, INGOs, and NGOs negotiate a division of labor during humanitarian disasters. For instance, one organization handles food aid, another handles WAter, Sanitation, and Health (WASH), and another curative health care (such as by setting up clinics in refugee camps). Violence prevention programs should be included on the list of essential services in volatile environments, and such programs should include preparedness support and, if needed, mobile aid.

An organization such as the Humanitarian Accountability Partnership (HAP—based in Geneva) should include early warning and early response in its programmatic scope. HAP has already incorporated conflict sensitivity of humanitarian aid into its portfolio, meaning that one measure of effectiveness of aid is how it impacts tensions within beneficiary populations.[16] Ethical principles and diagnostics should be developed to determine if humanitarian organizations are preparing for the worst, so that if field workers need to leave the area the aid beneficiaries are prepared to identify threats, negotiate if need be, and hibernate or flee to safety if they must. Those beneficiaries, as well, must know how to survive after fleeing, requiring, in many locations, a dispersed aid delivery capacity.

As more local-level violence prevention programs are developed and funded, evaluations will need to be conducted to help us learn from our mistakes. Here are some questions that can be asked in assessing successes and failures:

• Does the program develop "actionable" information for people who will be affected directly by the violence?

• Is early response "infrastructure" (meaning webs of relationships) in place at all three levels of leadership?

• Does the conflict early warning and early response system have the capacity to make order out of chaos? Does it provide coherent reports that are useful at each level of leadership?

• Does the early warning and early response network train people at the local level how to intervene to prevent violence and how to hibernate or flee to safety when prevention does not seem possible?

• Are people working in the field supported adequately? How do their resources compare to those being used by the headquarters?

• Do those working in the field have a protocol to follow when they are threatened? Is there a central phone number for them to call, a hibernation plan with provisions, and an evacuation plan?

• Are there redundant forms of communication? If cell phone towers are destroyed, for instance, how might the early warning and early response system be adapted quickly?

• Are there multiple types of data that are integrated in overlapping ways? Can people in conflict zones readily compare and contrast different types of information, such as the location of militant groups and the most recent provocative events, so they can make a comprehensive, informed determination of the level of threat?

• Does the approach being used have the right ratio of training people within target populations and providing preparedness support, to the amount of money and staff time being devoted to collect and analyze information at headquarters?

• Is technology helping local communities or is it distracting or endangering them?

• Given that limited resources are available, what information collection can be achieved through crowdsourcing rather through field monitors? Do field monitors need a permanent presence, or is it best that they focus on training leaders in one community after another, keeping them informed and integrated in a network of people with whom they communicate regarding violence prevention?

Conclusion

Innovation is all well and good, but you can easily miss the forest for the trees. The assault of visualization tools, of real-time, of data mining, of crowdsourcing . . . ; all these obscure the fact that our ability to respond to crises does not appear to improve much, year after year. The reason is simple: political problems cannot be solved by technological solutions, and at root most problems in ICT4Peace are political in one way or another.

—Paul Currion, "Conclusion"

Currion's assessment that politics trumps technology is a sobering reminder that we must always keep the big—including the political—picture in mind. But this does not detract from the main argument of this book: that violence prevention initiatives at a local level, combined with the support of middle- and top-level leaders, using various combinations of technology, have saved and can save lives. The question is not so much whether technology can be helpful, but what configuration is best in a given circumstance in view of limited financial and human resources constrained by security considerations of both staff members and the affected population.

This book stresses that violence prevention at a local level is an integral part of strategic peacebuilding in overcoming oscillations over time that plague societies which go into and out of periods of extreme violence. It offers an "applied theory of violence prevention" for reflective practice that has these dimensions: underlying conditions, time and space considerations, pathological social-psychological processes, and "levers" for impact. Of the underlying conditions, it highlights a history of enmity, a precipitating event, and norms.

Of the time and space considerations, the book focuses on the length of time between a precipitating event and the onset of violence—known as the lull—and seeks to address how long people have to intervene when warning signs are recognized after a precipitating event. It argues that those intervening to prevent violence need to be in close proximity to the group threatening attack in order to have a chance to counteract consensus building for violence.

The consensus-building pathologies involve emotional engagement, justification for killing, exaggeration of threat, reduction of risk, the lack of options, exploitation of grievances by militant leaders (the *greed* side of the *greed-grievance nexus*), and a desire to maintain honor and to fulfill a sacred duty. Each of these processes indicates that consensus building is taking place and, hence, there is a need to take action quickly to counteract it.

Among the levers used in taking action are outreach (by those with whom the group can relate—such as CeaseFire's approach to engaging former gang members), community engagement, community education, moral leadership, dialogue involving hermeneutic variability (often involving a struggle to overcome cognitive dissonance), engagement of police officers or peacekeepers, or both, and prosecution of perpetrators (especially of troublemakers and spoilers).

This book argues that there must be a balance between advocating for structural change and engaging in operational violence prevention. It covers the underlying analytical frameworks and theories and the various methodologies that are employed to help people not kill each other. It has distilled the criticisms of early warning and early response initiatives, involving concerns about resource allocation and failing to support vulnerable communities in knowing what to do when intervention to prevent violence is unlikely to work.

This book offers a preliminary finding of the length of the lull between a precipitating event and the outbreak of violence, reporting that, based on FCE's data, it is an average of 2.88 days. Sixty percent of the violence following a precipitating event occurred within two days. Over 75 percent of the violence was within the first four days. This adds credence to the view that supporting local capacity for local-level violence prevention is essential since marshaling interventions exclusively among outsiders is likely to be too late.

This book presents evidence that conflict early warning and early response at a local level has an impact on violence. It does not, of course, prevent it completely, but it does mitigate it. St. Xavier's had success preventing violence, though tensions surrounding religious symbols were more difficult to deescalate than those related to strictly secular matters. The case of CeaseFire offers solid evidence that the program reduced violence, both in neighborhoods and in hospitals. With FCE, there is statistically significant evidence that efforts to prevent violence reduced the number of human deaths compared to when interventions were not made. Similarly, with CEWARN, we found that efforts to prevent violence reduced the incidence of organized cattle raids. With Ushahidi, we saw that there is considerable promise in the use of crowdsourcing and digital mapping in responding to post-election violence. There is already evidence from the response to the earthquake in Haiti that crowdsourcing and digital mapping can save human life following a disaster. These technologies are likely to be most effective in conflict early warning and early response when combined with building trust networks, community organizing, bounded crowdfeeding, and restricted crowd feeding at grassroots, middle-, and top-levels of leadership so that early action can be initiated in locations where tensions are acute. Indeed, this is how social media was used in Tunisia and Egypt in the Arab Spring.

Violence prevention at a local level is essential to the success of strategic nonviolence. Not only is preventing the escalation of violence and counteracting its resurgence important in averting ethnoreligious violence, gang violence, post-election violence, and pastoral violence. It is also critical in overcoming the messy, difficult, overwhelming challenges involved in the transformation from oppressive dictatorship to democratic governance.

The five case studies of prototypical approaches to local conflict early warning and early response covered in this book provide the "raw material" for distilling common themes and propositions about the technologies, strategies, and tactics of nonviolence. The book argues that it is important to recognize the limits of trend analysis, pattern recognition, and visualization as compared to human induction. In fact, what is likely to be most effective is the combination of technology and induction. As the human mind has a propensity to hypothesize during a puzzle- or problem-solving quest, data—preferably visualized—can be interrogated, thereby sharpening the picture of what is happening or likely to happen. A person in an area of tension, seeing information from different sources, can infer meaning in ways that computers cannot, but computers can be helpful in pulling disparate information together, showing trends, identifying patterns, and providing summary pictures of complex situations. It is not necessary to put all eggs into one basket—either the human induction basket or the data gathering, analysis, visualization, and interrogation basket.

This book concludes that computerized pattern recognition can be a useful hedge against pathologies in information processing, whether due to individual biases or group "collective intelligence" malfunctions. It argues that there are cost efficiencies in using automated events-data generation software and crowdsourcing approaches. Yet there are advantages to having field officers—members of trust networks who consist of paid staff or volunteers, or a combination of the two—who report events on a regular basis and are available, if need be, for rapid verification of event reports and to intervene in tense situations on short notice, working side by side with local CBOs. Many societies have indigenous methods for the formation of peace-focused CBOs (such as a council of elders). These and other CBOs must be the nucleus of any local-level violence prevention approach.

This book argues that while events data need to be checked for validity, it is important to reconcile goals of relevance or utility, on the one hand, and time and accuracy, on the other, to determine when the identification of a trend is "good enough," because waiting for greater accuracy, however desirable, could prove catastrophic. It points out that this is for conflict early warning and early response what economists call a "second-best" solution. In other words, "the best should not be the enemy of the good."

In the five cases were example after example of how organizations limit the number of people to whom warnings are sent. Despite assertions of transparency, warnings typically are issued only to select individuals, with the possible exception of crowd feeding, though it is likely that even this will be limited to bounded crowd feeding or restricted crowd feeding, due mainly to concerns about causing panic and hysteria if warnings are disseminated widely.

This book provides examples of how to intervene effectively to prevent violence, relating them to the analytical frameworks and theories presented in chapter 1. The examples highlight the importance of deescalating emotional arousal, removing the justification for violence, increasing the sense of risk so that people know they will face consequences (such as going to jail) if they become violent, decreasing the sense of threat, creating viable options, and changing the norms that make violence an acceptable part of life. The book stresses, however, that it is important to plan for failure; that transnational organizations, INGOs, and NGOs need to prepare vulnerable communities for the worst. Preparedness of local communities should arguably be the highest priority in violence prevention programming, just as disaster preparedness is in flood-prone communities. Those at risk should be able to sustain themselves when transnational organizations, INGOs, and NGOs leave the area due to security concerns, and early warning is just as important for evacuation or hibernation as it is for early response.

Finally, this book projects how local conflict early warning and early response is likely to evolve. There is a need to exploit economies of scale, to develop ways to generate data through automation in languages other than English, and to explore the possibility of creating decision support software for determining when to issue warnings. It recommends that UN organizations, INGOs, and NGOs consider devoting a percentage of their resources to local conflict early warning and early response, and preparedness support, in volatile settings.

This book has not, however, covered some of the *critical components* to successful violence prevention capacity at a local level. These are the intangibles that cannot be paid for by donors, nor programmed by computer experts. And what I write here in the final paragraphs of this book might sound trite to some. But I would be remiss to not convey the importance of certain qualities of those who exercise leadership to effectively prevent violence at a local level. They need to have *credibility* with those who are threatened. They need to be *skilled at negotiation and persuasion*. At least some of them need to have *forceful personalities or charisma*, or both, so that when a warning is issued people will follow a leader to take action, whether that be to call in nearby outsiders with authority, get in the way of members of their group to protect those of another group, or to mediate, hibernate, or evacuate. And the local leaders, as well as those at the middle- and top-levels, must have *compassion*, a certain amount of *moxie*, and, not infrequently, considerable *courage*.

As this book comes to a close, I am reminded of those I have encountered in Asia, the Balkans, and Africa who put their lives on the line to intervene and prevent violence. They are an inspiration to me. Those who do this sort of work are not recruited in the ways of military recruiters. But they, at times, must be no less courageous than soldiers headed to a battlefield. Violence prevention practitioners rely on their "webs of relationships" and negotiation skills, walking into situations marked with uncertainty, thanklessness, and danger, without the assurance, however false that might be, of a gun in their hands.

Appendix A: Reporting Sheet for Field Officers[1]

Sending Date and Time: _____
Report Number: _____
Each report you submit needs a unique number. The first report you submit with this reporting sheet is number 1, the second number 2, and so on. Precede the number with the first letter of the district from which you are reporting.
Officer ID: _____ *Is this information sensitive? YES □ NO □*
Important: Please fill in all cells, if the necessary information is not available, indicate so.

1	*Sending Date and Time:*	
2	*Actual Date of Event:*	
3	*Event (WHAT):* *As a reference please use the list with events types*	
4	*Location (WHERE):* *Where did the event take place?*	
5	*Initiator (WHO):*	
6	*Recipient (TARGET):*	
7	*Information Source:* *From whom did you get your information?*	
8	*Information Credibility:* *How credible is the information (credible, ambiguous, questionable)?*	

9	Event Salience: Is the reported event relevant for the subnational, national, or international context?	
10	Injuries: How many people got physically injured through the event?	
11	Deaths: How many people died through the event?	
12	Material Damage: Describe what was damaged and the approximate value.	
13	Event Description: Short description of the event. All relevant information has to be included.	
14	Personal Observation: What could be the consequences of this event? Does it pose an immediate threat or a good opportunity for reconciliation?	
15	Early Response: Based on your personal observation, do you think taking actions to respond to this should be considered? If yes, what do you think should be done? Who can be contacted? Have you made these contacts or do you need staff members at Headquarters to do so?	
16	Link with Previous Events: Is this event linked with previous events? If yes, please indicate the Report Number of the previous event/s.	

Note

1. This reporting sheet was developed by numerous staff members of the Foundation for Co-Existence and Dominic Senn of swisspeace.

Appendix B: Categories for Local Conflict Early Warning and Early Response

This local conflict early warning and early response system was conceptualized in Colombo with FCE field officers and headquarters-based staff members during a workshop in January 2005 conducted by Dominic Senn of swisspeace and this book's author. Please note that Sri Lanka had suffered a tsunami, so the design incorporates conflict sensitivity during humanitarian relief operations.

Hostile Relationship:

- Perceived or actual access to justice that favors one group over another
- Perceived or actual access to education that favors one group over another
- Perceived or actual access to immigration (visas) that favors one group over another
- Perceived or actual access to jobs that favors one group over another, including aid agencies employing more people from one group than from other groups
- Perceived or actual provision of humanitarian assistance favoring one group over another
- Perceived ethnically motivated interruption of humanitarian assistance (such as a roadblock)
- Implicit ethical messages that create the perception that one group is more important or valuable than another group
- National policies that discriminate between different ethnic groups/different affected communities
- Theft of aid goods, especially by those connected directly to a warring side
- Implicit ethical messages that give legitimacy to violence-promoting leaders (troublemakers and spoilers) that bestow upon them more power, prestige, or access to international attention or wealth than is desirable (It

is unethical for outsiders to make troublemakers and spoilers appear to have authority when that has not been granted through legitimate political means by the local population.)

• Aid distributed in ways that legitimize the actions of war (for example, reinforcing patterns of population movements that warriors are causing)

• Aid distributed in ways that legitimize war-supporting attitudes (for example, rewarding those who are most violent; aid being given separately to all groups under the assumption that they cannot work together)

• Aid agency staff expressing discouragement and powerlessness in relation to their staff superiors, home offices, or donors, expressing disrespect for these people but often citing them as the reason why something is "impossible"

• Aid agency's publicity or fundraising approach or both demonizing one group, regularly treating another group as "victimized" by the other

Unacceptable Aid Agency Practices:

• Aid agency relying on arms to protect its goods or workers or both

• Aid agency refusing to cooperate or share information and planning functions with other aid agencies, local government, or local NGOs; openly criticizing the ways that others provide aid; and encouraging local people to avoid working with other agencies

• Field staff separating themselves from the local people with whom they are working, frequently using aid goods, or the power they derive from them, for their personal benefit or pleasure

• Aid agency apportioning its institutional benefits (salaries or per diem scales; equipment such as cars, phones, offices; expectations of time commitments to the job; rewards for work done; vacation, R & R, evacuation plans) in ways that favor one identifiable group of workers more than others

"Promotive" and Preventive Actions Taken to Correct Hostile Relationships:

• Advocacy for justice—for a more equal distribution of humanitarian assistance

• Formation of interethnic committees to determine aid allocations based on need

• Encouraging NGOs to use multiethnic teams

Precipitating Events:

• Physical violence and crime

• Perceived ethnoreligiously motivated killing

• Perceived ethnoreligiously motivated beating

- Perceived ethnoreligiously motivated raping
- Looting by one group at the expense of another

Symbolic Acts:

- Perceived ethnoreligiously motivated destruction by one group to a place of high symbolic value for another group
- Perceived ethnoreligiously motivated desecration of a holy site
- Perceived ethnoreligiously motivated interruption of a holy procession
- Perceived ethnoreligiously motivated damage to a sacred book, shrine, or place of worship

Changes in the Macrosituation:

- Political developments that are viewed as threatening to one group
- Substantial change in the status quo

Indications of a Lull:

- Reduction of constructive interaction between groups
- Reduction of cooperative acts between groups
- Discontinuance of meetings of inter-ethnoreligious formal associations
- Politicians, civic leaders, or religious leaders raising the profile of a precipitating event to a high level of visibility in a divisive fashion
- Arrival of new outside troublemakers or spoilers (on behalf of militant organizations or individuals)
- Activation of youth organizations/political troublemakers or spoilers
- Rumors

Acute Emotional Engagement:

- Group shouting or invading personal space of members of another group (such as yelling in very close proximity)
- Warnings given
- Threats made
- Protest campaigns
- "Hartals," such as protests, that are acts of aggression (as compared to those that are commemorative)

Note: The three categories that follow have various modes of communication: rumors, statements/speeches, handbills/brochures, newspapers, radio, television, posters/banners, and graffiti. They should be categorized as being communicated by one of these types of leaders: political, civic (such as business leaders or Rotary Club president), or religious.

Exaggeration of Threat Posed by the Out-group:

• Rumors of an imminent attack by one group against another

• Rumors of profound economic loss that will result from one group maintaining a presence

• Rumors of a profound reduction of current or anticipated humanitarian assistance that will result from one group maintaining a presence

Reduction of the Risk of Violence to the In-group:

• Statements made by politicians, the police, or the army that those who kill will not be brought to justice (will not go to jail, will not be prosecuted)

• Rituals designed to protect future attackers (such as invulnerability practices in which a shaman bestows protection onto warriors before battle, allegedly making them immune from harm)

Justification for Killing:

• Interpretation of a precipitating event as a moral violation of one group against another

• Statement made by a religious leader that the faithful have a duty to "teach" the other group "a lesson"

Positive Developments/Cooperation:

• Aid agency has sought to identify intergroup bridge-building activities in the conflict area that cross the boundaries and connect people on different sides and has designed its program to relate to these connectors

• Aid delivered in ways that reinforce a local sense of inclusiveness and intergroup fairness; programs designed to bring people together so that for any group to gain, all groups must gain

• Aid delivered in ways that reinforce, rather than undermine, attitudes of acceptance, understanding, and empathy between groups

• Aid delivered in ways that provide opportunities for people to act and speak in as they did before the violence

• Aid agency providing opportunities for its local staff to cross lines and work with people from the "other" side

• Aid respecting and reinforcing local leaders as they take on responsibility for civilian governance

• Aid providing rewards for individuals, groups, and communities that take intergroup or peace-reinforcing initiatives

• Aid agency staff reinforcing attitudes of their friends and counterparts as they remember, or reassert, sympathy and respect for other groups

Event Linkage:

- Is this event linked to another event (give serial number)?

Prognosis:

- How likely is violence? Explain.
- If there is violence, how soon will it occur? Explain.

Early Response Actions Taken to Preempt Violence:

- Counteract rumors/myth busting (fact finding, putting a precipitating event into context, challenging exaggerations)
- Request for moderate leaders to intervene to prevent violence (especially with mid- and top-level leaders), making it clear publicly that there will be consequences for becoming violent (the perpetrators will be brought to justice)
- Appeal to the key tenets of a given religion that all human beings are of value (statements against tribalism and xenophobic religious nationalism)
- Assertively but respectfully engage with troublemakers and spoilers, in a fashion that allows them to save face publicly
- Undertake public confrontation of troublemakers and spoilers only when efforts to neutralize their influence privately have failed
- Provide safe haven for groups in imminent danger

Outcome of Actions Taken:

_____

_____

Contact information of moderate influential leaders in this location (local-, middle-, and top-level leaders):

_____

_____

Contact information of moderate influential leaders outside this location who might be able to influence local leaders:

_____

_____

Appendix C: "Super Event" Categories

PROVOCATIONS =

- Main Event—Force, Subevents: Vehicle bombing, Sexual assault, Beatings, Assassination, Coups and mutinies
- Main Event—Other incident, Subevent: Missing
- Main Event—Request, Subevent: Call for action, Solicit support
- Main Event—Accuse, Subevent: Other accuse, Criticize or denounce
- Main Event—Threaten, Subevents: Threat forceful occupation, Threat forceful attack, Threaten to boycott or embargo
- Main Event—Threaten, Subevent: Give ultimatum
- Main Event—Demonstrate, Subevent: Other protest demonstration, Other demonstration, Protest defacement, Protest obstructions, Protest procession, Armed force activation
- Main Event—Seize, Subevent: Hostage taking & kidnapping, Other abductions, Other arrest & detention, Attempt abduction, Political arrests

MITIGATION =

- Main Event—Consult, Subevents: Engage in negotiation, Mediate talks, Other consult, Other discussions, Travel to meet, FCE intervention
- Main Event—Force, Subevent: Crowd control
- Main Event—Grant, Subevent: Evacuate victims
- Main Event—Agree, Subevent: Agree to peacekeeping, Agree to settlement
- Main Event—Propose, Subevent: Offer to mediate, Offer to negotiate

VIOLENCE =

- Main Event—Force, Subevents: Other force use, Other physical assault, Other bombing, Other conventional weapon attack, Riot or political turmoil, Small arms attack, Suicide bombing, Artillery attack, Armed battle, Other armed actions

- Main Event—Human conditions, Subevent: Human death
- Main Event—Election activities, Subevent: Election violence
- Main Event—Seize, Subevent: Other bombing
- Number of human deaths
- Number of human injuries
- Monetary value of damage to property

Other "super events" that we created but did not use in our analysis, but that might be helpful for others using this data base, include:

INITIATIVES =

- Main Event—Endorse (all Subevents)
- Main Event—Promise (all Subevents)
- Main Event—Grant, Subevents: Other provide shelters, Other ease actions, Relax administrative sanctions, Relax censorship, Relax curfew, Ease economic sanction, Grant asylum
- Main Event—Agree, Subevents: Other agree or accept, Agree to mediation, Agree to negotiate
- Main Event—Request, Subevent: Request mediation
- Main Event—Propose, Subevents: Other propose, Offer peace proposal

EXCHANGES =

- Main Event—Consult, Subevent: Host a meeting
- Main Event—Other incident, Subevent: A & E performance, Sports contest
- Main Event—Grant, Subevent: Other release or return, Return release persons, Release return property, Improve relations, Extend invitation
- Main Event—Agree, Subevent: Collaborate

AID =

- Main Event—Reward, Subevent: Extend economic aid, Extend humanitarian aid, Extend military aid, Other reward

WARNINGS =

- Main Event—Warn, Subevent: Armed force alert, Other alert, Other armed force display, Other warn, Security alert

AGGREVATORS =

- Main Event—Comment, Subevent: Pessimistic

- Main Event—Force, Subevent: Torture, bodily punishment

- Main Event—Request, Subevent: Request an investigation

- Main Event—Reject, Subevent: Disclose information, Break law, Impose censorship, Impose restrictions, Other refuse to allow, Other reject, Other reject proposal, Other defy norms, Political fight, Reject ceasefire, Reject mediate, Reject peacekeeping, Reject proposal to meet, Reject request for material aid, Reject settlement

- Main Event—Complain, Subevent: Other complain, Formally complain, Informally complain

- Main Event—Deny, Subevent: Deny

- Main Event—Threaten, Subevent: Threaten to halt mediation, Threaten to halt negotiations, Threaten to reduce or break relations, Threaten to reduce or stop aid, Other sanctions threat, Other physical force threats, Other armed force threats, Other threaten, Nonspecific threats

- Main Event—Sanctions, Subevent: Other sanctions, Other reduce or stop aid, Other halt discussions, Strike & boycott, Break relations, Halt mediation, Halt negotiations

- Main Event—Expel, Subevent: Expel

- Main Event—Seize, Subevent: Covert monitoring, Other seize, Other seize possession

- Main Event—Election activities, Subevent: Election activities

PRESSURE =

- Main Event—Other incident, Subevents: Extreme climatic condition, Flood, Hazardous material spill, Hurricane, Drought, Earthquake, Animal death, Animal illness, Volcano, Wildfire, Tsunami, Tornado, Other natural disaster

- Main Event—Economic activity, Subevent: Government default on payment

- Main Event—Human conditions, Subevents: Equity prices down, Other human illness

CONSENSUS BUILDING FOR VIOLENCE =

- Main Event—Human conditions, Subevent: Tension

- Main Event—Rumors, Subevents: Conflict rumor, Other rumor, Preventive rumors

Appendix D: Indicators of the CEWARN Mechanism[1]

1. Description of CEWARN Indicators: Situation Reports

Alliance Formation

- Interethnic group alliance
- Ethnic group—government alliance

Armed Intervention

- Internal armed support
- External armed support

Behavioral Aggravators

- Interrupt other activities
- Pastoral migration
- Bullets as commodities
- Development aid problems
- Harmful migration policy
- Protest
- Media controls
- Harmful livestock policy
- Student attendance interrupted
- Migrant laborers
- Influx of IDPs
- Separation of groups
- New markets
- Security escorts
- Livestock prices dropped
- Negative media coverage

- Small arms availability
- Post-raid blessing
- Livestock sales increase

Environmental Pressure

- Natural disaster
- Land competition
- More livestock in secure areas
- Grazing areas abandoned
- Livestock disease

Exchange Behavior

- Celebration
- Intergroup marriage
- Gift offering
- Intergroup sharing
- Cross-border trade

Mitigating Behavior

- Access to health care
- Relief distributions
- Law enforcement
- Small arms disclosure
- Markets remain open
- Bride price stable
- Access to education
- Positive media coverage
- Negotiations taking place

Peace Initiatives

- Women peace messengers
- Weapons reduction program
- Local peace initiatives
- Religious peacebuilding
- NGO peace initiatives

Triggering Behavior

- All-male migration
- Pre-raid blessing
- Traditional forecasting

2. Description of CEWARN Indicators: Incident Reports

Armed Clashes

- *Military Battle* (Armed hostilities or engagements between an official military unit of a government and an armed party. Includes both civil war and interstate war battles.)
- *Other Armed Clashes* (All other armed hostilities or engagements. Includes all communal and intercommunal battles.)

Raids

- *Raids with Abductions* (Raids focused on abductions of people or the taking of hostages. May include injuries or death to humans, and/or damage, destruction, or theft of other property.)
- *Organized Raids* (Other organized raids. May include injuries or deaths to humans, and/or damage, destruction, or theft of other property.)
- *Livestock Theft* (Raids focused on the theft of livestock. May include injuries or death to humans, and/or damage, destruction, or theft of other property.)

Protest Demonstrations

- *Peaceful Protests* (Peaceful protest demonstrations or assemblies which may include isolated or low-level violence.)
- *Violent Turmoil or Riots* (Assemblies or crowds that get out of control marked by violence, disorder, damage, and/or destruction.)

Other Crime

- *Assaults* (Physical attacks and abuse involving the actual use of physical force against individuals, and/or groups. Does not include abductions.)
- *Banditry* (Commandeering of vehicles, highway robbery, and other similar criminal activities.)

Note

1. Taken, with minor editorial changes, from http://www.gtz.de/de/dokumente/en-CEWARN-Strategy-2006.pdf, pp. 40–41, accessed October 17, 2011.

Appendix E: Results from Statistical Analysis on Organized Raids

Using the data from Meier, Bond, and Bond 2007, here are the results from our regression analysis with de-trended data on organized raids

	Estimate	Std. error	t value	p
(Intercept)	2.980	4.977	0.599	0.550
Alliances	-0.073	0.060	-1.226	0.222
Exchanges	-0.142	0.066	-2.158	0.033
Mitigation	0.216	0.071	3.040	<0.01
Initiatives	-0.107	0.065	-1.652	0.101
Aggravators	0.415	0.135	3.076	<0.01
Pressure	-0.102	0.054	-1.887	0.061
Provocation	-0.117	0.075	-1.549	0.124
Rainfall	0.064	0.054	1.181	0.240
Vegetation	0.0263	0.026	1.003	0.318
Forage	-0.003	0.003	-0.952	0.343

Acronyms

AMIS   African Union Mission in Sudan-Darfur
AMM   Africa Media Monitor
AOR   area of responsibility
ATCC   Association of Peasant Workers of Carrara
ASF   African Standby Force
BJP   Bharatiya Janata Party
CADE   Center for the Advancement of Distance Education
CBO   community-based organization
CSO   civil society organization
CEWARN   Conflict Early Warning and Response Mechanism
CEWERU   Conflict Early Warning and Early Response Unit
CEWS   African Union's Continental Early Warning System
CIA   Central Intelligence Agency
CICR   Center for International Conflict Resolution
CIDA   Canadian International Development Agency
CPMR   Conflict Prevention, Management, and Resolution
CRS   Catholic Relief Services
DRC   Democratic Republic of Congo
EAWARN   Network for Ethnic Monitoring and Early Warning
ECHO   European Commission's Humanitarian Aid Office
ECOWAS   Economic Community of West African States
EMM   European Media Monitor
EROS   Eelam Revolutionary Organizers
FARC   Armed Revolutionary Front of Colombia
FASE   Fuzzy Analysis of Statistical Evidence
FAST   a German acronym for "Early Analysis of Tensions and Fact-finding"
FBR   Free Burma Rangers
FCE   Foundation for Co-Existence
FEWER   Forum on Early Warning and Early Response

FIRST   Facts on International Relations and Security Trends
FM   field monitor
FORECITE   Forecasting of Crises and Instability using Text-based Events
GIS   geographic information system
GPS   Global Positioning System
GI-NET   Genocide Intervention Network
GoSL   Government of Sri Lanka
GTZ   German Agency for Technical Cooperation
HMM   hidden Markov models
IBO   Internet Bar Organization
ICG   International Crisis Group
ICT   information and communications technology
ICT4Peace   Information and Communication Technologies for Peace
IDEA   Integrated Data for Events Analysis
IDPs   internally displaced people
IED   improvised explosive device
IFRCRC   International Federation of Red Cross and Red Crescent Societies
IGAD   Inter-Governmental Authority on Development
INGO   international non-governmental organization
IPPP   International Peace and Prosperity Project
IT   information technology
JRC   European Commission's Joint Research Center
KEDS   Kansas Events Data System
LAN   local area network
LTTE   Liberation Tigers of Tamil Eelam
MACD   moving average convergence divergence
MFOA   Media Focus on Africa Foundation
MIT   Massachusetts Institute of Technology
NGO   non-governmental organization
NRI   National Research Institute
OCHA   United Nations Office for the Coordination of Humanitarian Affairs
OECD   Organisation for Economic Co-operation and Development
OPP   Orangi Pilot Project
OSCE   Organization of Security and Co-operation in Europe
PLOT   People's Liberation Organization of Tamil-Eelan
PSC   Peace and Security Council
RFID   radio-frequency identification
RRF   Rapid Response Fund

RSS   Really Simple Syndication
RTC   Responding to Conflict
R2P   responsibility to protect
SDC   Swiss Agency for Development and Cooperation
Sida   Swedish International Development Cooperation Agency
SIM   subscriber identity module
SIPRI   Stockholm International Peace Research Institute
SLA   Sri Lanka Army
SLMC   Sri Lanka Muslim Congress
SLMM   Sri Lanka Monitoring Mission
SMS   Short Message Service
TAF   The Asia Foundation
UNHCR   United Nations High Commissioner for Refugees
UNITAR   United Nations Institute for Training and Research
UN MONOC   United Nations Mission to the Democratic Republic of Congo
USAID   United States Agency for International Development
V&TCs   volunteer and technical communities
VRA   Virtual Research Associates
WANEP   West Africa Network for Peacebuilding
WEIS   World Events Interaction Survey

Glossary

Accelerator dimensions of intergroup relations that serve to escalate tension, often developing a momentum of their own (Gurr and Harff 1996).

Actionable information data, sometimes depicted graphically, in the case of early warning with which one can determine if violence is likely in order to decide whether to intervene, hibernate, or flee.

Agenda-setting function when a social marketing campaign influences the attention of a community, thereby impacting public discourse (McCombs and Shaw 1972).

Agent-based early warning system an early warning approach that analyzes the statements of one or more pivotal "agents" or leaders (O'Brien 2010). Such systems also often use fundamental and technical indicators.

Al-Mi'rage the belief among Muslims that Mohammad the prophet ascended to heaven from Jerusalem.

Arab Spring pro-democracy and anti-dictatorship movements in the Middle East that began in late 2010, mirroring "European Spring" liberation movements in Central and Eastern Europe following the downfall of the Soviet Union.

Automated early warning system an approach for violence prevention using events data generated through a computer program.

***Badal* code** the tradition of blood revenge common among ethnic Pukhtoons in South Asia.

Bayesian modeling an approach whereby clusters of words are used to establish categories that are honed by developing and refining hypotheses of the likelihood of association. The modeling software, once developed, will look for "clues" in each report to determine the most likely event category.

Bounded crowd feeding sending warnings of potential violence to a core group of trained people who can intervene to prevent violence and are capable of being prudent with sharing information that might otherwise cause panic, disseminated also to part of the crowd in close proximity to an outbreak of potential violence.

Bounded crowdsourcing a way of enhancing the accuracy of information by limiting sources to a trusted group of reporters. Combining this method with open crowdsourcing is a hybrid approach (Heinzelman and Waters 2010, 12).

Calibration focus groups gatherings of local experts who assign weights to indicators of conflict and cooperation.

Civil society organizations volunteer organizations at a local level.

Cognitive dissonance theory a theory developed initially by Festinger (1957) which postulates that when a person holds thoughts that are in a dissonant state, the cognitive discomfort results in unconscious or conscious processes seeking to obtain consonance.

Communalism a label used in South Asia, sometimes called "ethnoreligious nationalism," which combines religious passion with political ambition parochially, as if one ethnoreligious group is superior to others.

Communal harmony a state of inter-ethnoreligious relations in which different groups embrace diversity rather than trying to marginalize those who are different.

Community-based organizations small non-governmental organizations that operate in a local area and nowhere else, usually operating with volunteer members and officers.

Community organizing facilitating the formation of civil society entities for purposes of social action and development, often considered a subfield of social work.

Conflict early warning and early response an approach which seeks to identify when and where violence is likely to happen so that it may be prevented.

Conflict management efforts to contain and transform violent conflict once it erupts.

Conflict resolution initiatives designed to resolve conflict, assuming that conflict is a challenge that can be overcome.

Conflict transformation a process that does not necessarily bring resolution of conflict but that at least develops approaches for dealing with it constructively, assuming that tension is part of human relations and that conflict can be channeled in positive directions.

Consensus-building process the social psychological processes that occur during a "lull." According to Horowitz (2001), these processes include (1) acute emotional engagement involving anger, arousal, rage, outrage, or wrath; (2) exaggeration of threat posed by the out-group; (3) reduction of the risk of violence to the in-group; and (4) justification for killing (sometimes using religion to indicate that a moral violation has occurred).

Crowd feeding   sending alerts about potential violence to observers of events who have participated in data generation, often based on a combination of data validation and pattern recognition.

Crowd seeding   distributing cell phones to people who agree to be field monitors.

Crowdsourcing   a form of events data collection that allows volunteer observers to identify and report events. This term was introduced by Baker (2006).

Crowdsourcing honesty   an approach whereby people are reminded of the morality of submitting accurate information (Ariely 2008, 212–213).

Crowdsourcing the filter   a method being developed by Ushahidi of involving the crowd in assessing the validity of data sent from others in the crowd whereby events are assigned values of green (for accurate), yellow (for unknown), and red (for inaccurate). As more people in the crowd assign values to events posted on a digital map, the scores are added together and a determination is made on the accuracy of the report. Events determined to be inaccurate are deleted from the digital map.

Complex emergencies   those that involve violent conflict and humanitarian disasters (such as hurricanes, floods, famine, drought, and epidemics).

Conflict early response   the goal of conflict early warning, which is to facilitate a response early enough to prevent violence.

Conflict early warning   the field of study and practice of using indicators of conflict and cooperation to identify situations of acute turmoil and tension that could lead to violence.

Conflict transformation   a process that does not necessarily bring resolution of conflict but that develops approaches for dealing with inevitable tensions constructively.

Data harvesting   extracting information from social media and other Internet-based sources either manually or by using automation.

Data mining   searching data and using algorithms to identify patterns in it.

Diffusion of responsibility   a concept developed initially by Darley and Latané (1968) to explain why people do not intervene to help a victim who is being assaulted as long as there are many people witnessing the violence.

Digital straw   a mathematical technique developed by staff members and volunteers of Ushahidi in which data being sent from a noisy environment are siphoned based on redundancy, proximity, and credibility before being placed onto a digital map.

Equity theory   a group of social science theories that hold that perceptions about inequities of land, job opportunities, and the like, mechanistically result in violence.

Events data   information about events of cooperation and conflict that are coded.

Expert system   "A special program that resembles a collection of 'if . . . then' rules. The rules usually represent knowledge contained by a domain expert (such as a physician adept at diagnosis) and can be used to simulate how a human expert would perform a task" (Flake 1998, 451).

Extremist leaders   people with credibility in civic, political, or religious circles, or a combination thereof, who command respect of their in-group and are prone to exclusion and violence. They are sometimes called "militant leaders," though a distinction can be made between militancy and violence.

Flashpoint asymmetry problem   when there has been a history of enmity and a precipitating event, the rate at which intergroup goodwill can be depleted is an order of magnitude faster than the rate at which intergroup goodwill can be created.

Free-rider problem   a concept developed initially by Olson (1971) whereby collective goods are enjoyed by those who do not help to bring them about.

Fundamental indicators   measures that are relatively stagnant that change gradually, such as demographic pressure and the extent of human rights violations (also called *structural indicators*).

Fuzzy logic   a capacity to identify associations of words as well as different meanings of the same word, used in artificial intelligence applications (Zadeh 1965; King and Lowe 2003, 638).

Grievance   what a group perceives as discrimination, prejudice, or injustice.

Greed-grievance nexus   a phrase used to describe how violence can be promoted when some are calling attention to inequity and injustice (grievances) while at the same time others who have something to gain from violence (power, prestige, pecuniary gain) are fanning the flames of discontent using hate speech, politically sophisticated and manipulative symbol choreography, and other forms of incitement.

Hartel   a term used in South Asia for a strike of an extreme nature involving a virtually complete shutdown of businesses and shops, often accompanied by violence.

Hidden Markov models   a mathematical tool of analysis, developed initially by Andrey Andreyevich Markov in 1921, for generating probabilistic predictions of moving from one state to another. For instance, one state might be "low intensity conflict" while another is "massive armed conflict." These models are hidden to the extent that they are not directly observable. For instance, one cannot tell whether or not a conflict is massive by seeing an isolated event of fighting.

Incident reports   documentation of episodes of violence when they occur.

Information fire hose   an onslaught of data which is increasing in both volume and speed.

Iniquity theory   an encapsulated way of explaining how religious leaders with credibility within a faith community can artfully appeal to a sense of honor and sacred duty, urging followers to become violent when facing an offense to what is deemed sacred (developed by Clark McCauley in Bock and McCauley 2003).

Interest-based negotiation   an approach to mediation that entails helping groups in tension to identify their interests rather than their positions.

Locally led advance mobile aid   building local capacity to identify threats of imminent violence so local inhabitants are prepared and can take action themselves to intervene to prevent violence, hibernate, or flee—while also being able to survive when in hibernation or after fleeing, as with IDPs (internally displaced people) who are vulnerable to a hostile belligerent military or paramilitary group.

Lull   the period between a precipitating event and violence during which a consensus is developed to become violent.

Mashup   a website that aggregates data in useful ways. It is a term coined by Erik Hersman (Ekine 2009, xxii).

Moral dissonance   an uncomfortable psychological state a person feels when doing harm to others.

Non-governmental organizations   nonprofit entities that usually have paid staff members who operate in multiple communities.

Operational codes   an axiomatic articulation of decision processes developed by identifying and analyzing statements made by key individuals (such as a commander of a paramilitary force) or small groups of people (such as leaders of a political party). This approach is sometimes used as part of an *agent-based early warning system*.

Operational violence prevention   intervening to prevent violence by using such techniques as mediation, making arrests, or persuading a belligerent group not to attack.

Paradox of early warning   the earlier a warning is, the less compelling a threat seems (Barrs 2004).

Parsing   to separate words in a sentence into categories, often associated with generating automated events data.

Pattern recognition   a way of identifying something based on what we can discern now for the purpose of making a decision, often involving movements, graphs, or mathematics, or a combination.

Precipitating event   an event in the past or anticipated in the future that has considerable symbolic potency, causing groups of people to begin developing a consensus for violence.

Predictive tagging analyzing components of a crowdsourced report to determine if another report is of the same event, thereby preventing multiple counting of a single event while adding credibility to the data.

Preparedness support a specialized form of disaster preparedness that builds the capacity of vulnerable people to detect a threat of violence, mediate, hibernate, or flee from it and survive.

Protection a term used widely within humanitarian circles and among international lawyers, meaning that international organizations should protect vulnerable people, such as those who are displaced within their own countries—both IDPs and refugees. A priority usually is placed on the protection of women and children. So, for instance, the installation of lights near latrines used by women and children in a refugee camp is considered part of "protection services" because the presence of light is expected to reduce the amount of rape and child molestation. Because international law tends to focus on protection of refugees and not IDPs owing to notions about state sovereignty (that countries should stay out of the internal affairs of other countries), some international organizations have sought to increase the sense of responsibility of governments for their own populations. In 2006, U.N. Security Council Resolution 1674 was passed, which places a responsibility to protect (R2P) on governments to prevent "genocide, war crimes, ethnic cleansing and crimes against humanity and the international community's responsibility to assist states in fulfilling their responsibility" (Asia-Pacific Centre for the Responsibility to Protect 2009: 1).

Reality score a measure of accuracy of events data based on four mathematical approaches—"peer review," reliability, proximity, and redundancy—developed by Ushahidi.

Release theory an encapsulation of views regarding collective violence that holds that catharsis occurs when people act on their hostile impulses toward an individual or group.

Religious illiteracy "the low level or virtual absence of second-order moral reflection and basic theological knowledge among religious actors— . . . a structural condition that increases the likelihood of collective violence in crisis situations [which creates a condition whereby] . . . A supremely self-interested and skilled politician or preacher confronting—or having assembled—a mass of people outraged by their 'victimization' at the hands of ethnic or religious others may easily exploit deep emotional currents and volatile prejudices [involving] . . . The selective retrieval and politically motivated interpretation of one dramatic episode from a vast, complex, and ambiguous history . . . [that] serves to construct and demonize 'the other,' to solidify and channel extremist passions, and to extend a sacred canopy over the whole dubious process" (Appleby 2001, 69).

Response report descriptions of interventions, or lack thereof, that cover what was done following a warning, who did what, and whether the violence prevention effort resulted in success or failure.

Restricted feeding sending warnings only to select individuals, mainly trained volunteers and staff members, along with trusted officials, if applicable, primarily due to concerns that warnings can cause panic and hysteria. This is distinct from *bounded crowd feeding* or *crowd feeding* in that they involve part or all of "the crowd," respectively.

Risk analysis of countries ranks "fragility" on the basis of various "fundamental" indicators such as the extent of discrimination, unemployment of the young and educated population, and land tenure.

Shared awareness "the ability of many different people and groups to understand a situation, and to understand who else has the same understanding" (Shirky 2008, 163).

Short-circuiting proposition a view that holds that if one or more of the components in Horowitz's (2001) analytical framework is counteracted through various interventions, then violence will be averted. This involves: counteracting emotions of arousal, rage, outrage, or wrath; challenging a view that killing is justified; reducing the sense of threat; and increasing the sense of risk of becoming violent.

Spoilers those who disrupt peace processes out of boredom, thrill seeking, feeling left out, or another self-serving purpose.

Spoilers to stakeholders strategy a way of implementing reconstruction initiatives that makes those disrupting the peace process have a stake in the end of violence.

Social media technology that "relates to sharing of information, experiences, and perspectives throughout community-oriented websites . . . [involving] blogs, forums, message boards, picture- and video-sharing sites, user-generated sites, wikis, and podcasts" (Weinberg 2009, 1).

Sons of the soil conflicts intergroup tension and violence related to a belief by one group that it has sacred entitlement to a specific place and a duty or religious mandate to hold on to the land.

Strategic nonviolence an approach to conflict transformation designed to bring about political change by increasing, rather than decreasing, societal tensions, but with little or no bloodshed, requiring considerable organizing, strategy, and leadership.

Strategic peacebuilding "an approach to reducing violence, resolving conflict and building peace that is marked by a heightened awareness of and skillful adaptation to the complex and shifting material, geopolitical, economic, and cultural realities of our increasingly globalized and interdependent world. Accordingly, peacebuilding that is strategic draws intentionally and shrewdly on the overlapping and imperfectly coordinated presences, activities and resources of various international, transnational, national, regional, and local institutions, agencies and movements that influence the causes, expressions, and outcome of conflict. Strategic peacebuilding takes advantage of emerging and established patterns of collaboration and interdependence for the purposes of reducing violence and alleviating the root causes of deadly conflict" (Lederach and Appleby 2010, 22).

Structural indicators   measures that are relatively stagnant that change gradually, such as demographic pressure and the extent of human rights violations (also called *fundamental indicators*).

Structural violence prevention   changing local, national, and international policies and procedures that result in less tension between different identity groups.

Situation reports   conveyed information about developments in intergroup relations (both of conflict and cooperation) often involving trend analyses, but sometimes including warnings that violence is likely to happen.

Technical indicators   measures that are dynamic, that can change dramatically within a short time frame.

Theory of second best   a methodological convention that affirms that data cannot be 100 percent reliable if it is to be timely and that reasonably reliable data are satisfactory. The theory of second best was developed by Lipsey and Lancaster (1956) in economics. It is sometimes called non-ideal theory. It relates to an environment of imperfect information, holding essentially that "the accurate should not be the enemy of the good."

Troublemakers   those who promote the use of violence to overcome challenges of injustice, to achieve domination, or to gain power, prestige, or wealth in some other way.

Trust network   a group of people collaborating in a volunteer and technical community who are dedicated to a cause and share mutual trust.

Volunteer and technical communities (V&TCs)   associations comprised of core groups of ITC specialists—part social movement, part NGO—which creates a platform to enable people to take action.

Warning-response problem   an observed deficiency that early warnings often have not resulted in early responses.

Xenophobic religious nationalism   a condition existing in situations in which leaders of a group use religion to claim superiority of their group relative to other groups, marked by close-mindedness to other perspectives.

Notes

Preface

1. The "middle-out" idea was first developed by Lederach (1997).

2. The graphic developed by Chris Blow makes this point. It shows his estimation of the allocation of time involved in getting a crowdsourcing platform like Ushahidi set up as compared to other required activities to make an initiative successful. See his blog post at http://blog.ushahidi.com/index.php/2010/05/19/allocation-of-time-deploying-ushahidi/, accessed October 31, 2011.

3. On the background of international youth movements working together, and on U.S. diplomatic initiatives urging Mubarak to resign, see David D. Kirkpatrick and David E. Sanger, "A Tunisian-Egyptian Link That Shook Arab History," *The New York Times*, February 13, 2011, http://www.nytimes.com/2011/02/14/world/middleeast/14egypt-tunisia-protests.html?pagewanted=all, accessed October 14, 2011.

4. For information on the use of the Ushahidi crowdsourcing platform by Egyptian activists, see http://blog.ushahidi.com/index.php/2011/02/03/egypt-ushahidi-jan25/, accessed October 29, 2011.

5. The website is at www.aeinstein.org, accessed October 27, 2011.

6. The website is www.ushahidi.com, accessed October 27, 2011.

Introduction

1. Zarrella (2009) provides a somewhat different definition: "Social media comes in many forms . . . the eight most popular [are]: blogs, microblogs (Twitter), social networks (Facebook, Linkedin), media-sharing sites (YouTube, Flickr), social bookmarking and voting sites (Digg, Reddit), review sites (Yelp), forums, and virtual worlds (Second Life)."

2. Shirky calls this phenomenon *shared awareness* which he defines as "the ability of many different people and groups to understand a situation, and to understand who else has the same understanding" (2008, 163).

3. Note that these would be considered by many to be risk analysis systems rather than early warning systems, but I do not see the value of such a distinction since many of these approaches are now being combined.

4. Again, some of these would be viewed as risk analysis rather than early warning systems. While the Global Events Data System is identified as being part of the first generation, it did, in fact, have field monitors that collected quantitative events data, making it a second-generation system (Davies 1998). I am indebted to Joe Bond for pointing this out.

5. "Fourth Generation Early Warning and Early Response," blog posting, March 6, 2009, http://earlywarning.wordpress.com/2009/03/06/fourth-generation-early -warning-systems/, accessed October 14, 2011.

6. This point is made well by David Carment in an e-mail communication relative to postings on preventing-conflict@googlegroups.com with Patrick Meier, accessed March 6, 2009.

7. For more information on this project, see http://www.satsentinel.org/about, accessed April 9, 2011. I am indebted to Patrick Meier for pointing this project out to me.

8. For information on Voix des Kivus, go to http://cu-csds.org/projects/event-mapping-in-congo/. For an illuminating presentation about it by Peter van dor Windt, go http://crisismappers.net/main/authorization/signIn?target=http%3A%2F %2Fcrisismappers.net%2Fvideo%2Ficcm-2009-voix-des-kivus. Both sites accessed January 30, 2011.

9. This type of technology can, of course, be used for other purposes. A colleague working in Gaza Strip informs me that it is being used, accompanied with sophisticated weaponry, for targeted assassinations.

10. For additional information on the conflict early warning and early response approach used by the Canadian International Institute of Applied Negotiation, go to http://www.ciian.org/erunit1.shtml, accessed November 2, 2011.

11. For information on the history and methodology of the Grameen Bank and the work of Mohammad Yunus, see http://www.grameenfoundation.org/who-we-are, accessed October 14, 2011.

12. "The Economics of Conflict," DECRG Ongoing Research on the Economics of Conflict, The World Bank, http://econ.worldbank.org/WBSITE/EXTERNAL/EXTDEC/ EXTRESEARCH/EXTPROGRAMS/EXTCONFLICT/0,,menuPK:477971~pagePK:64168 176~piPK:64168140~theSitePK:477960,00.html, accessed August 1, 2008.

13. For definitions of "durable" and "stable peace" at a macrolevel, see Lund (1996, 39). This perspective that violence prevention is strategic in a post-conflict—in addition to a pre-conflict—situation differs from the view offered by Woocher (2009),

but his focus is at a macrolevel, analyzing what Nyheim (2008) has referred to as first- and second-generation early warning systems. Here, instead, I am focusing on third- and fourth-generation systems involving local conflict, using events data produced by field officers or by using a crowdsourcing approach.

14. British Broadcasting Network, Sri Lanka Country Profile, http://news.bbc.co .uk/2/hi/south_asia/country_profiles/1168427.stm, accessed October 28, 2011.

15. "A Micro Level Analysis of Violent Conflict," MICROCON, http://www .microconflict.eu/aboutus.html, accessed July 1, 2008. Boldface in original.

1 Toward an Applied Theory of Violence Prevention

1. John Paul Lederach, in Mindanao, the Philippines, October 21–26, 2007, speaking at a workshop on peacebuilding with staff members of Catholic Relief Services and partner organizations.

2. It should be noted that Chigas also cautioned that strong "bonding social capital" can also result in increased violence. The question is, how accurate is the information and the number and effectiveness of moderate leaders?

3. Stewart provides a helpful analysis of the differences between ethnic and religious conflict, acknowledging how the two identities often overlap while clarifying important distinctions (2009).

4. Note that this is a variation of the perspective offered initially by Simon and Schon (1996). See also Hatchuel 2002.

2 Reporting and Warning about Deadly Possibilities

1. For more information on fundamental and technical indicators used by stock traders, see *Investor's Business Daily* at http://www.investors.com, accessed June 26, 2009.

2. swisspeace lost funding for the FAST program. Therefore, it is not possible to provide a citation to it. swisspeace has numerous other projects, however, that may be of interest to readers. See www.swisspeace.ch, accessed October 15, 2011.

3. In a World Bank publication (2011, 28), the authors contend that the international community should move "from sporadic early warning to continued risk assessment." This can be confusing to those interested in local-level early warning. Keep in mind, however, that experience has shown that early warning systems for the international community have had disappointing results due mainly to bureaucratic and political decision processes and, one can argue, physical distance. In that light, one can see this recommendation as an acknowledgment of the difficulty of marshaling early response at the international level (using "first generation" or "second generation" systems).

4. Integrated Data for Events Analysis, http://www.vranet.com/idea/, accessed April 14, 2009.

5. Phone conversation with Joe Bond, March 1, 2011.

6. In an email communication with Damindra De Silva on June 29, 2009, I learned that the Foundation for Co-Existence is already at an experimental stage of using handheld devices to ask people to assign weights to categories.

7. Of those, data on harassment and assassinations, for instance, could be made independent variables. And ethnic violence would be, for example, a dependent variable. The relationships of the two independent variables to the dependent variable can be determined. One can then develop weights according to the amount explained by one variable compared to another in a regression model. Note that another way to assign weights is to do "principal component analysis," which examines data being generated by the system and reduces the numbers of variables while retaining the information of the original variables. This was brought to my attention by Dennis Culhane.

8. For more information on the "digital straw," see http://irevolution.net/2009/04/10/developing-swift-river-to-validate-crowdsourcing/, accessed October 15, 2011.

9. Explaining Swift River, http://blog.ushahidi.com/index.php/2009/04/09/explaining-swift-river/, and Crisis Info: Crowdsourcing the Filter, http://blog.ushahidi.com/index.php/2009/02/04/crisis-info-crowdsourcing-the-filter/, accessed October 15, 2011.

10. This type of approach was used by Avencia (now called Azavea) in its Crime Spike Detector (now called "Hunchlab"), though the simple formula was modified so events reported minimally do not produce misleading warnings. For information on Azavea's approach, go to http://www.azavea.com/, accessed February 25, 2011.

11. Conversation with Joe Bond, International Studies Association convention, San Francisco, CA, March 27, 2008.

II Violence Prevention on the Ground

1. Thomas W. Malone, http://cci.mit.edu/about/MaloneLaunchRemarks.html, accessed February 26, 2009.

3 Organizing against Ethnoreligious Violence in Ahmedabad

1. I acknowledge with appreciation the generous hospitality of Fr. Cedric Prakash and his dedicated staff members during and following my field research (including assistance with translations from Gujarati to English), and the guidance and support I received from Mary B. Anderson, then president of Collaborative Development Action in Cambridge, MA.

2. Other explanations of the work of St. Xavier's Social Services Society can be found in Bock 1999, also available electronically at http://www.developmentinpractice .org/journals/communal-conflict-ngos-and-power-religious-symbols, accessed October 15, 2011, and published in an abbreviated form in Mathews 2001, 22–23.

3. One such instance was in the Maghaninager area of Ahmedabad.

4. The district collector is the head administrator of a district in the Indian political system. The holder of this position also has the power of executive magistrate, which can be invoked during times of civil disturbances. In that capacity, this office holder has policing authority greater than a municipal police commissioner.

5. Examples include: (1) In November and December 1990, St. Xavier's provided medical assistance to people hurt during rioting from a pool of nearly ten thousand who were living in temporary relief camps. CRS/Bombay provided $5,000 for this relief operation. (2) In January 1993, following the December 1992 riots, St. Xavier's distributed 800 tins of edible oil, 1,300 blankets and 500 kilograms of wheat flour, benefiting some 5,184 people. In addition, St. Xavier's, with a grant from Campana of Spain, provided an average of two thousand rupees per family (roughly $67) as a grant to help people purchase lumber and other building materials for the reconstruction of their homes. Roughly 800 homes were rebuilt with these grants. People also were given money from the government for this purpose.

6. For instance, during a flood emergency in July 1993, when Ahmedabad faced its worst flood since 1973, St. Xavier's, along with the Missionaries of Charity Sisters, set up medical clinics to respond to outbreaks of cholera and intestinal illness due to the unsanitary conditions resulting from the high water. Plastic sheeting, bamboo, medical supplies, and food were provided, mainly in the Shahpur Darwaja and Shantipura areas where nine feet of water had washed people's houses away, and in the Khanjiara Darwaja (which is one of the twenty outreach slums) and Saptarushi Badhar areas close to Haveli. Here, 2,000 people assembled in a nearby municipal school for shelter. These activities were funded by CRS/Bombay and by Aide A L'enfance d L'inde (Luxembourg), each providing a separate grant of around $8,333.

7. In 1992, for example, St. Xavier's provided growth-monitoring services for 701 children in Sankalitnagar, 161 children in Mahajan-no-Vando, and 199 children in Nagori Kabarasthan; monitored prenatal patients of which 225 were in Sankalitnagar, 55 in Mahajan-no-Vando, and 76 in Nagori Kabarasthan; conducted eight "health camps" throughout Ahmedabad (one of which was in Sankalitnagar and another in Mahajan-no-Vando), one aimed at teenage girls and two others on the health of women; and conducted five month-long training programs for community health workers. In 1993, St. Xavier's provided community health services through its clinics and outreach workers to 8,649 patients in the urban slum settlements of Sankalitnagar (5,441 patients), Mahajan-no-Vando (1,810 patients), and Nagori Kabarasthan (1,398 patients); 254 prenatal checkups, 537 growth monitoring visits, and 1,104 immunizations in Sankalitnagar; 61 antenatal checkups, 102

growth monitoring visits, and 179 immunizations in Mahajan-no-Vando; and 53 antenatal checkups, 150 growth monitoring visits, and 228 immunizations in Nagori Kabarasthan; midwife training for ten of its village health workers; a testing and certification program for participating village health workers, who were provided six workshops on naturopathy, homeopathy, and accupressure (St. Xavier's attempts to promote alternative medical approaches); six health education meetings in various slums, three on tuberculosis, two on general health practices, and one on ante- and postnatal health; diagnosis and treatment to 173 tuberculosis patients in the slums; and community health education in response to a jaundice epidemic in the slums, during which close to two hundred people in Ahmedabad died of the illness (none of the areas in which St. Xavier's had a health program were affected).

8. The fact that Indian law tends to protect slum dwellers from being evicted may have the effect, however unintentional, of increasing the manipulative cultivation of riots by real estate developers. Apparently, if the slum dwellers' property is completely torn down and abandoned temporarily during a riot, their legal rights relative to preventing eviction are jeopardized. In such cases, a person from each concerned family is often asked to go back to the violence-stricken area to "hold on to the plot" at considerable physical risk.

9. Interview with Fr. Cedric Prakash, Ahmedabad, India, August 24, 1994.

10. Allah and Ishwar are both labels for God in Arabic and Sanskrit, respectively.

11. Various art forms are an integral part of building and maintaining peace. According to the European Centre for Conflict Prevention, quoting the actress Baisa Baki from war-torn Bosnia, "Theatre has another meaning for us. It is not a luxury any more, but a necessity. During the war it was the only way to give shape to our thoughts, our pain, the nightmare. That doesn't mean we only played tragedies. On the contrary, there was a great need for comedies. To be free from suffering for a moment. Now we are trying to reconstruct our identity by looking for plays by Bosnian dramatists. We will fill the holes in our culture" (1999, 159).

12. I would be remiss to not recognize that safe havens have sometimes been places of mass execution, as in Rwanda.

13. I first heard the phrase "getting in the way" when interacting with the Christian peacemaker teams in Hebron, West Bank. See http://www.youtube.com/user/CPT Hebron, accessed July 10, 2011.

14. There is a legend behind the exchange of rakhis. According to it there was once a Hindu queen who was being attacked. She needed some strength from a Muslim ruler whom she asked for protection. When the Muslim ruler consented, the queen tied a rakhis around the Muslim's wrist.

15. "St. Xavier's Social Services Society, Report 1992–1993," 16.

16. It is noteworthy that the local police are trained in mediation. In fact, sometimes the police commissioner calls a meeting of community leaders in an effort to preempt conflict. Interview with K. N. Shelat, district collector, Ahmedabad, January 3, 1995.

17. In a statement emphasizing the importance of St. Xavier's learning from its mistakes, Prakash wrote that St. Xavier's realizes "how difficult it is to promote true harmony at the grass-roots level." "St. Xavier's Social Services Society, Report 1992–1993," 7.

18. It is important to make a distinction here between *fundamentalism*, which denotes having strong beliefs based on the literal translation of a holy book, and *religious militancy*. Whether the existence of the former increases the propensity for the latter, of course, depends upon a host of factors and conditions.

19. "St. Xavier's Social Service Society, Report 1992–1993," 7.

20. This assumes, of course, that the police are professional in the sense that they do not take sides and become a belligerent force themselves.

21. For a theoretical discussion about approaches used relative to the various levels, see DeMars 1993.

4 Overcoming Gang Violence in Chicago

1. Quote from remarks given at a National Institute of Justice conference in 2009, http://www.ojp.usdoj.gov, accessed October 15, 2011.

2. For more information on Second Life, go to http://secondlife.com/, accessed December 6, 2010.

3. To see more about CADE's work with CeaseFire, go to http://rwjfblogs.typepad .com/pioneer/2008/12/more-from-candice-kane-of-ceasefire.html, accessed October 15, 2011.

4. http://rwjfblogs.typepad.com/pioneer/2008/12/more-from-candice-kane-of -ceasefire.html, accessed October 28, 2011.

5. Quote from http://www.nij.gov/journals/264/ceasefire-interrupters.htm, accessed October 15, 2011.

6. For a helpful article on the role of religious leaders, see Ritta M. Basu's "Interupting Violence" in Duke Divinity School's *Faith & Leadership* site, http://www .faithandleadership.com/features/articles/interrupting-violence?page=0,0, accessed October 29, 2011.

7. Taken from http://www.ceasefirechicago.org/clergy.shtml, accessed November 19, 2010. See also http://ceasefirechicago.org/education/chicago-youth-and-clergy -members-sit-down-at-the-peace-table, accessed October 15, 2011.

8. Phone conversations with Colleen A. Monahan, director of CADE. For more information see http://www.advancedrealities.com/docs/CeaseFire_Island_Violence _Prevention.pdf, accessed October 15, 2011.

9. Note that an earlier three-year CeaseFire program was implemented in South Bend under the auspices of Notre Dame's Robinson Center, but it ran into difficulties over confusion relating to the provision of job opportunities and for-profit objectives, on the one hand, and nonprofit, violence prevention efforts, on the other. One of the biggest challenges of dealing with gang violence is in providing constructive alternatives to the members, especially help in getting a job. For information on SchoolTipline, go to http://www.schooltipline.com, accessed March 24, 2011.

10. This figure is taken directly from Skogan et al. 2008, I-4. For a video on the background of this theory of change, see http://www.poptech.org/popcasts/gary _slutkin__poptech_2008, accessed October 15, 2011.

11. Quoted from http://www.ceasefirechicago.org/DOJ%20study.shtml, accessed November 19, 2010.

12. Quote from page 2 of the Brief Summary found at http://www.northwestern .edu/ipr/publications/ceasefire.html, accessed November 19, 2010.

13. http://www.advancedrealities.com/docs/CeaseFire_Island_Violence_Prevention .pdf, accessed October 28, 2011.

14. I had a number of conversations about the use of the Second Life site with Moynihan during 2009 and 2010. See http://www.ojp.usdoj.gov/nij/journals/264/ ceasefire.htm#author, accessed October 15, 2011.

15. This quote is from http://blog.ushahidi.com/index.php/2010/10/29/peacetxt/, and is also found at http://www.poptech.org/peacetxt, both accessed October 15, 2011.

5   Counteracting Ethnoreligious Violence in Sri Lanka

1. This project has resulted in *Paradise in Tears*, a book by Victor Ivan. See http:// www.thesundayleader.lk/2009/12/20/paradise-in-tears-%E2%80%93-new-edition -by-victor-ivan/, accessed October 20, 2011.

2. Figures given in a presentation by Kumar Rupesinghe, president of FCE, at a workshop entitled "Early Warning Systems: Potential for Crisis Management and Regional Cooperation," organized by the University of Karachi, Department of International Relations, and the Hanns Seidel Foundation, Karachi, Pakistan, March 18, 2009.

3. Other organizations that intervene in tense situations to prevent conflict include UNICEF, ICRC, the Local Monitoring Committee (LMC), and SLMM. Interview with K. M. H. Akbar, Local Monitoring Committee, Trincomalee, August 12, 2004.

4. We bifurcated the type of violence by using data on *initiators* and *recipients* of events of conflict and cooperation. These initiators, listed verbatim from a "drop-down" menu on FCE's events data entry system, were categorized as events between groups of different religious identities: Muslims, Tamils, Sinhalese, Buddhists, People's Liberation Organization of Tamil-Eelan (PLOT), Civilians, Religious Agents, Ethnic Agents Organization, Ethnic Actors, Eelam Revolutionary Organizers (EROS), and the Sri Lanka Muslim Congress (SLMC). And these recipients of cooperation or conflict (also listed verbatim from another of FCE's drop-down menus) were likewise categorized as events of groups of different religious identities: Ethnic Agents Organization, Ethnic Groups, Muslims, PLOT, Religious Agents, Sinhalese, SLMC, and Tamils. Only those events that had one of these initiators *and* recipients were then labeled as events that were not related directly to the larger separatist revolt. In other words, for instance, if a Government of Sri Lanka military force engaged in conflict or cooperation with an ethnic group, we categorized that as an event related to the larger separatist revolt, not between groups of different religious identities. During the major fighting of the separatist revolt, the Government of Sri Lanka was seen by many as a Buddhist, Sinhalese force. While it is correct that many in the government were both, not all were.

5. I am indebted to Stephanie Croos for her hard work in reviewing PROVOCATION data.

6. For this clarification, I benefited from Meier, Bond, and Bond 2007.

7. Using a t-test, we found that this is a statistically significant difference at the 0.05 level.

8. Of the 2,476 measurements of the lull, only 38 were greater than 100 (1.5 percent of the cases). The standard deviation with all these data points is 42.78. If we truncate using three times the standard deviation we eliminate anything above 128. This yields a median of 1 day, a mean of 4.842 days, and a maximum of 124 days. In contrast, when we eliminate the 38 cases above 100 days, the standard deviation drops to 9.17. This difference shows how outliers can have such an effect on the mean and in turn the standard deviation. So our removal of outliers included the 38 cases above 100 (which we believe is conservative) and then anything above three standard deviations from the mean, or 27. It is our assessment that anything over 27 days either relates to recurring violence or the precipitating event and the violence are unrelated.

9. We derive a 95 percent confidence interval of 2.69–3.06.

6   Crowdsourcing during Post-election Violence in Kenya

1. This historical account of linking geographical information to events data and how this capacity might impact early warning and early response was derived from "Crimson Hexagon: Early Warning 2.0?"—an excellent blog entry by Patrick Meier

found at http://earlywarning.wordpress.com/2009/02/17/crimson-hexagon-early-warning-20/, accessed March 10, 2009. Readers can also find information about Crimson Hexagon at http://www.crimsonhexagon.com, accessed on October 16, 2011, and WarViews at http://www.icr.ethz.ch/research/warviews, accessed on March 29, 2009. For more information, see the blog post "A Brief History of Crisis Mapping," http://irevolution.net/2009/03/12/a-brief-history-of-crisis-mapping/, accessed October 20, 2011.

2. Quotation taken from http://www.icr.ethz.ch/research/warviews, accessed March 29, 2009.

3. From http://www.ushahidi.com, accessed March 20, 2009.

4. See http://irevolution.net/2009/04/09/ushahidi-comes-to-india-for-the-elections/ accessed on October 16, 2011. To see the website on the Democratic Republic of Congo, go to http://drc.ushahidi.com/?lang=en_US, accessed May 9, 2009.

5. Using cell phones to send messages condemning violence can be an effective way to prevent escalation, though we are just beginning to understand what is most effective in what context. An example of when cell phone messaging seems to have been useful is when the chief executive officer of a cell phone company in Kenya, Safaricom, who was asked to turn off the network to prevent hate messages, decided to do the opposite. He sent messages of peace, and warned that anyone using text messaging to propagate hate would be prosecuted. See http://irevolution.net/, accessed October 16, 2011.

6. See http://blog.ushahidi.com/index.php/2008/12/01/thoughts-on-hot-flash-conflict-in-mumbai-and-nigeria/, accessed May 9, 2009.

7. The Model Code of Conduct can be found at http://eci.nic.in/eci_main/faq/faq_mcc.pdf, accessed October 16, 2011.

8. For more information on Vote Report India, go to http://blog.ushahidi.com/index.php/2009/04/07/vote-report-india-launches/, accessed October 16, 2011.

9. The Vote India website is found at http://votereport.in/, accessed April 29, 2009.

10. Using Bayesian methods to develop categories is but one of the software components Ushahidi is developing. See http://appfrica.com/2008/09/05/ushahidi-relanches-website-secures-new-funding/, accessed February 16, 2011. For more on Bayesian methods, see Earman 1992 and Tenenbaum 1999.

11. Information provided by Patrick Meier, April 9, 2011.

12. A story about this lock defect—Leander Kahney's "Twist a Pen, Open a Lock"—is available at http://www.wired.com/culture/lifestyle/news/2004/09/64987, accessed April 6, 2011.

13. See http://irevolution.net/2009/04/09/ushahidi-comes-to-india-for-the-elections/ and http://irevolution.net/2009/01/02/crowdsourcing-honesty/, accessed October 16, 2011.

14. This is explained by Patrick Meier at http://irevolution.net/category/ushahidi/ page/3/, accessed October 16, 2011. Note that Meier uses the color orange instead of yellow in his example. Meier also indicates that during the Haiti earthquake response a feature called "digg" was added, which essentially provides an opportunity for those in the crowd to rate the accuracy of a report. Communication with Patrick Meier on April 4, 2011.

15. Patrick Meier pointed this out. See http://irevolution.net/2009/05/07/moving-forward-with-swift-river/, accessed October 16, 2011.

16. Patrick Meier pointed this out to me, April 9, 2011.

17. See Patrick Meier's blog post, "Moving Forward with Swift River," at http://irevolution.net/2009/05/07/moving-forward-with-swift-river/, accessed October 16, 2011.

18. Ibid.

19. See http://blog.ushahidi.com/index.php/2008/11/05/sms-reporting-through -ushahidi/, accessed March 3, 2009.

20. Harvesting data using predictive tagging, combined with these approaches to check validity of data quickly in near real time, collectively comprise Ushahidi's Swift River platform component. For more information, see http://blog.ushahidi .com/index.php/2009/04/09/explaining-swift-river/, accessed October 19, 2011.

21. See http://blog.ushahidi.com/index.php/2009/02/04/crisis-info-crowdsourcing -the-filter/, accessed May 9, 2009.

22. See http://blog.ushahidi.com/index.php/2008/12/01/thoughts-on-hot-flash -conflict-in-mumbai-and-nigeria/, accessed May 9, 2009.

23. Ibid.

24. This information was provided by Media Focus on Africa. See http://www .mediafocusonafrica.org/index.php?Division=Organisation, accessed February 24, 2011. An explanation of the Peace Heroes project can be found at Butterfly Works' website at http://www.butterflyworks.org/content/3802/media_focus_on_africa, accessed October 16, 2011.

25. For a description of efforts to reduce post-election violence, see the "Unsung Peace Heroes" post at http://www.peaceheroes.ushahidi.com/winners.php, accessed February 24, 2011.

26. See the July 3, 2010, blog post "Libya: Whose Information Are We Sharing?" at http://blog.ushahidi.com/index.php/2010/07/03/liberia-whose-information-are-we -sharing/, accessed November 29, 2010.

27. For a counterexample (pointed out to me by Ian Delrossa) of an effective advocacy initiative of Greenpeace that involved lots of communication but almost no action by the people doing the communicating, see this TED Talk: http://www.ted.com/talks/alexis_ohanian_how_to_make_a_splash_in_social_media.html, accessed April 6, 2011.

28. For more information on crowdsourcing warning and response, see http://earlywarning.wordpress.com/2008/10/17/crowdsourcing-warning-and-response/, accessed June 29, 2009.

29. The crisis map for Libya is available at http://libyacrisismap.net/bigmap. Background about OCHA's request and the team that put the map together is at http://blog.ushahidi.com/index.php/2011/03/06/using-new-ushahidi-map-libya/, both accessed March 7, 2011.

30. Paul Currion, "If All You Have Is a Hammer—How Useful Is Humanitarian Crowdsourcing?" *MobileActive.org*, posted October 20, 2010, http://mobileactive.org/how-useful-humanitarian-crowdsourcing, accessed July 8, 2011.

31. "Allocation of Time: Deploying Ushahidi," Ushahidi blog, posted on May 19, 2011, http://blog.ushahidi.com/index.php/2010/05/19/allocation-of-time-deploying-ushahidi/, accessed July 8, 2011.

32. Currion, see note 30.

33. "Evaluating Crowdsourcing for Humanitarian Response," *jungle light speed*, posting undated, http://www.junglelightspeed.com/evaluating-crowdsourcing/, accessed July 8, 2011.

34. I am indebted to Monica Palmeri for the information in this section, which she shared via email on January 12, 2011.

35. I first encountered the reference to "smart crowds" in a presentation at Harvard University at a summit convened by the Harvard Humanitarian Initiative in 2011. Much of what was presented at the summit can be found in Harvard Humanitarian Initiative (2011), also found at http://www.unfoundation.org/assets/pdf/disaster-relief-20-the.pdf, accessed October 16, 2011.

36. I am reminded of an example of synergism among technologies presented by Peter Senge. He points out how the first commercially and aerodynamically viable plane, the DC-3, was first introduced in 1935, thirty years after Orville and Wilbur Wright proved that powered flight was possible. McDonnell-Douglas, the corporation that made the DC-3, pulled five technologies together. According to Senge, they were "the variable pitch propeller, retractable landing gear, a type of light weight molded body construction called 'monocque,' a radial air-cooled engine, and wing flaps. To succeed, the DC-3 needed all five; four were not enough" (1996, 6).

37. "The Arab Spring & Western Policy Choices," *Peace Policy*, Kroc Institute for International Peace Studies, University of Notre Dame, posted July 6, 2011,

http://peacepolicy.nd.edu/2011/07/06/the-arab-spring-western-policy-choices/, accessed July 8, 2011. For an insightful analysis of the challenges facing proponents of democratic transition in the Middle East, see the article by Larry Diamond at http://liberationtechnology.stanford.edu/news/a_fourth_wave_or_a_false _start_20110523/, accessed July 11, 2011.

7 Circumventing Tribal Violence in East Africa

1. This section draws heavily on Meier, Bond, and Bond 2007.

2. For a description of the various districts and counties in this cluster as they relate to data used for this study, see Meier, Bond, and Bond 2007, 723. For more recent information on the geographical reach of CEWARN, go to http://cewarn.org/index. php?option=com_content&view=article&id=51&Itemid=53, accessed November 30, 2010.

3. For more information on the Rapid Response Fund, go to http://cewarn.org/ index.php?option=com_content&view=article&id=61&Itemid=94, accessed November 30, 2010.

4. Alerts are found at http://cewarn.org/index.php?option=com_docman &Itemid=98, accessed February 24, 2011.

5. A thoughtful reviewer of my manuscript, Joe Bond, explains: "These include algorithms used in market forecast such as Bollinger bands, the ultimate oscillator, moving average convergence/divergence, among others." The approach is similar to that used by O'Brien (2002). In a conversation with Joe Bond on March 1, 2011, he explained how VRA is working with three early warning and early response systems in Africa: CEWARN, the Early Warning and Response Network (ECOWARN), and the African Union's Continental Early Warning System (AU-CEWS). Bond points out that the data from situation and incident reports constitute independent variables, while the dependant variable is a measure of instability based on indicators developed by the Heidelberg Institute for International Conflict Research. For more information on this type of analysis, see O'Brien 2010.

6. Phone conversation with Joe Bond, March 1, 2011.

7. Quote from http://cewarn.org/index.php?option=com_content&view=article&id =97&Itemid=131, accessed November 30, 2010. Much of the information in this section is taken from this website.

8. Ibid.

9. Ibid. Community radio can, of course, be used to facilitate violence as well as prevent it. A report by the Center for International Media Assistance warns that "the power of community radio to mobilize groups and bring change to societies is well

recognized. This power can, however, also be manipulated and used to spread hate and violence, as was the case in Rwanda in 1994" (2007, 8).

10. For more information on Botswana's approach, go to http://practicalaction.org/ peace5_cattle_tracking_botswana, accessed December 2, 2010.

8   Comparing the Approaches

1. Definition from MedicineNet.com, http://www.medterms.com/script/main/art. asp?articlekey=10638, accessed June 20, 2009.

2. As the designers of the Crime Spike Detector note, their system has "no explanatory or interpretive power" and is "not good for low volume events" (Cheetham and McGinnis 2009, 19).

3. As an example of many people coming up with an accurate determination, Sunstein offers this example: "The British scientist Francis Galton sought to draw lessons about collective intelligence by examining a competition in which contestants attempted to judge the weight of a fat ox at a regional fair in England. The ox weighed 1,198 pounds; the average guess, from the 787 contestants, was 1,197 pounds" (2006, 24). Additional information for this section was derived from http:// en.wikipedia.org/wiki/The_Wisdom_of_Crowds, accessed April 9, 2009.

4. The Philadelphia Police Department has made some of its maps available to the public; see http://citymaps.phila.gov/CrimeMap/, accessed February 27, 2009.

5. See   http://irevolution.net/2009/03/27/internews-ushahidi-and-communication -in-crises/, accessed October 17, 2011.

6. Ibid.

9   How to Intervene Effectively

1. This is from an unpublished document sent via e-mail from Madhawa Palihapitiya in August 2007.

2. Here is where the research on types of violent conflict sheds light on an exception to this general rule. During genocides, a very specific kind of violence, members of the in-group who protect those of the out-group are killed. This is due, in part, to the mentality that everyone must participate in the killing and that "anyone not for us is against us."

3. Interview   found   at   http://www.justpeaceint.org/endingcyclesofviolence.php, accessed February 9, 2011.

4. Draft manuscript on preparedness training sent to me by Casey Barrs on May 14, 2009. Emphasis in original.

5. Interview by Julian Portilla, found at the website of the Beyond Intractability Knowledge Base Project, funded by the William and Flora Hewlett Foundation, http://peacestudies.beyondintractability.org/audio/10548/, accessed July 13, 2011.

6. Note that Lederach (2005, 13–16) also deserves credit for translating García 1996.

10   What to Do When Violence Prevention Is Unlikely to Work

1. There is a heated debate about this idea of "protection." Some argue that, realistically, there is very little capacity among UN agencies and almost no political will on the part of foreign governments to engage in protection, so well-intentioned statements on the "responsibility to protect" (also known as "R2P") are misguided and can inadvertently leave already vulnerable people even more vulnerable because of the false hopes thereby created. This view is well articulated by Wertman (2009).

2. Email communication with Casey Barrs, June 20, 2008.

3. This is consistent with observations gleaned by the Local Capacities for Peace Project; see Anderson (1999).

4. Email message from Casey Barrs to Ben Hoffman (copied to a number of others), March 17, 2009.

5. This section reflects substantial input from Casey Barrs, who generously provided feedback on an earlier draft.

6. Taken from http://www.freeburmarangers.org/About_Us/, accessed October 18, 2011.

7. See http://www.freeburmarangers.org/About_Us/, accessed May 9, 2009.

8. Email communication with Casey Barrs, May 14, 2008; phone conversation with Barrs, May 16, 2008; and phone conversation with Dave Eubank, June 24, 2009.

9. I want to be clear that what is written here should not be viewed as criticism of Free Burma Rangers. In fact, the more I learned about their work, the more impressed I was with their approach and dedication.

10. Our phone conversation was on June 24, 2009. I was humbled as I listened to Dave Eubank, reflecting later on what a fine example he and his family provided, especially juxtaposed to the challenging conversations my wife and I have had with our son regarding military service. We have always emphasized how being willing to risk one's life for others in need is what we are asked to do, but that, in the military, one has very little control over what one is ordered to do; that being willing to die is far different from being willing to kill. It gave me pause to wonder how many young males would opt for the Free Burma Rangers over, for instance, the U.S. Marine Corps.

11. For instance, staff security training on tactical response to threats, such as those now offered by the International Federation of Red Cross and Red Crescent Societies (IFRCRC) and European Commission's Humanitarian Office (ECHO), can be opened to civilian observers such as wardens in refugee camps.

12. Taken from http://www.genocideintervention.net/protection/projects, accessed May 9, 2009. The new site is at http://www.endgenocide.org/who-we-are, accessed October 18, 2011.

13. As quoted by Casey Barrs from Beah 2007 in his letter to Ben Hoffman, March 17, 2009.

11 Concerns about Misallocation of Resources

1. FORCITE uses the VRA Reader for automated events data generation from Reuter's News Briefs, similarly to other macro early warning systems. See Yiu and Mabey 2005, 43.

2. By "connectors" we mean those defined by Mary B. Anderson (1999) as linkages in a society that foster cooperation and local capacities for peace, compared to "dividers" that do the opposite, such as providing aid to one group at the expense of other groups in need.

3. Quoted from http://news.bbc.co.uk/2/hi/africa/3573229.stm, accessed June 20, 2009.

12 Future Directions and Recommendations

1. See http://rwjfblogs.typepad.com/pioneer/2008/05/the-second-life.html, accessed June 21, 2009.

2. On the use of communication technologies in countries were civil liberties are constrained, see http://irevolution.net/2009/06/15/digital-security/, accessed October 18, 2011.

3. This analysis relies heavily on Patrick Meier's post, "Crimson Hexagon: Early Warning 2.0?" at http://earlywarning.wordpress.com/2009/02/17/crimson-hexagon-early-warning-20/. See also http://www.crimsonhexagon.com/, both accessed October 18, 2011.

4. Damindra De Silva, senior manager of IT at the Foundation for Coexistence, introduced many of the ideas in this section to me during the summer of 2008.

5. For more information on Freedom Fone, go to http://www.netsquared.org/projects/freedom-fone, accessed June 29, 2009. I am indebted to Patrick Meier for drawing my attention to this website.

6. I refer here to the book *1984* by George Orwell (1949) in which privacy was compromised by totalitarian governmental surveillance.

7. Joe Bond made this point in a conversation on March 1, 2011.

8. For more information, see Meier's post and see Crimson Hexagon website, note 3, both accessed on March 10, 2009.

9. Quotation taken from Meier post, note 3, accessed March 29, 2009.

10. In Ghana, counterfeit drugs are being tracked similarly, though the system in use there does not analyze images. Instead, pharmaceutical customers send a text message with the serial number that is on the drugs' packaging. See The Economist 2011. I am grateful to Katie Persons for bringing this article to my attention.

11. Information on Spyglass can be found at http://cellphoneforums.net/apps/navigation-7/spyglass-ar-tracker-compass-sextant-gps-maps-pavel-ahafonau-24752/, accessed October 18, 2011.

12. For helpful background on developing this kind of approach, see Bazerman and Watkins 2004, 198.

13. In many cases, a p-value of 0.05 is considered on the borderline of acceptable, meaning that the result of the statistical test has a probability of being wrong 5 percent of the time.

14. It is noteworthy that the Philadelphia Police Department has found it beneficial to integrate other kinds of data, along with information on criminal events. This is similar to how Meier, Bond, and Bond (2007) found that analyses of pastoralist-pastoralist and agriculturalist-pastoralist violence in the Horn of Africa are improved considerably by integrating environmental indicators with conflict and cooperation events data.

15. For a brief explanation of the system dynamics approach, go to http://www.iiasa.ac.at/Research/POP/pde/htmldocs/system.html, accessed February 20, 2011.

16. A document written by Mary B. Anderson of CDA Collaborative Learning Projects entitled "Options for Aid in Conflict" is being used by HAP and can be found at http://www.hapinternational.org/pool/files/options-for-aid-in-conflict-pdf1.pdf, accessed June 30, 2009.

References

"Arab Spring Hardens into Summer of Stalemates as Challenge of Changing Regimes becomes Clearer." 2011. *The Washington Post,* July 14.

Ackleson, Jason. 2006. "Mapping Ethnic Violence." *International Studies Review* 8 (3): 492–494.

Adelman, Howard. 1999. "Early Warning and Ethnic Conflict Management: Rwanda and Kosova." *Conflict Trends* (3): 13–19.

Adelman, Howard. 2006. *The War Report—Warning and Response in West Africa.* Ghana: USAID/WARP Accra.

African Union. 2006. "Meeting the Challenge of Conflict Prevention in Africa—Towards the Operationalization of the Continental Early Warning System. Background Paper No. 1. Report of the Workshop on the Establishment of the AU Continental Early Warning System (CEWS)." Kempton Park, South Africa. December. http://www.google.com/url?sa=t&rct=j&q=&esrc=s&source=web&cd=1&ved=0 CBoQFjAA&url=http%3A%2F%2Fwww.africa-union.org%2Froot%2Fua%2FConfere nces%2Fdecembre%2FPSC%2F17-19%2520dec%2FRpt%252030-31Oct%2520%252 003%2520Background%2520n%25201.doc&ei=SZegTvvlMa2IsAKk08mHBQ&usg= AFQjCNF9Pg_NkCkfxXYpRHNsXTrjiqH3Gw&sig2=64jKu-gq6Uv66Y0j8DA52Q, accessed October 20, 2011.

Allison, Graham, and Philip Zelikow. 1999. *Essence of Decision: Explaining the Cuban Missile Crisis.* 2nd ed. New York: Longman.

Anderson, Mary B. 1999. *Do No Harm: How Aid Can Support Peace—Or War.* Boulder, CO: Lynne Rienner Publishers.

Annan, Kofi. 1999. *Toward a Culture of Prevention: Statements by the Secretary General of the United Nations.* New York: Carnegie Commission on Preventing Deadly Conflict.

Appleby, R. Scott. 2001. *The Ambivalence of the Sacred: Religion, Violence, and Reconciliation.* Lanham, MD: Rowman & Littlefield Publishers.

Ariely, Dan. 2008. *Predictably Irrational: The Hidden Forces That Shape Our Decisions.* 1st ed. New York: Harper.

Asia-Pacific Centre for the Responsibility to Protect. 2009. *The Responsibility to Protect and the Protection of Civilians: Asia-Pacific in the UN Security Council Update No. 1.* Brisbane, Australia: The University of Queensland.

Atallah, Amjad. 2004. "Planning Considerations for International Involvement in an Israeli Withdrawal from Palestinian Territory." *Journal of Humanitarian Assistance* (July 1): 1–28.

Atallah, Nabil. 1998. "Muslim Leaders Shaken by Al-Aqsa Pig Plot." *The Jerusalem Times,* January 2.

Austin, Alexander. 2004. *Early Warning and the Field: A Cargo Cult Science?* Berghof Handbook for Conflict Transformation. Berlin: Berghof Center. http://www.berghof -handbook.net/documents/publications/austin_handbook.pdf, accessed October 20, 2011.

Austin, G., E. Brusset, M. Chalmers, and J. Pierce. 2004. *Evaluation of the Conflict Prevention Pools—Synthesis Report.* London: Department for International Development.

Ayers, Michael D. 2003. Comparing Collective Identity in Online and Offline Feminist Activists. In *Cyberactivism: Online Activism in Theory and Practice,* ed. Martha McCaughey and Michael D. Ayers, 145–164. New York: Routledge.

Baker, Chris. 2006. "Q&A in Second Life: Crowdsourcing and Virtual Worlds." *wired. com,* November 8. http://www.wired.com/gamelife/2006/11/qa_in_second_li/, accessed October 20, 2011.

Banks, Ken. 2011. Appropriate Mobile Technologies: Is Grassroots Empowerment for All? In *Mobile Technologies for Conflict Management: Online Dispute Resolution, Governance, Participation,* ed. Marta Poblet, 27–38. New York: Springer.

Barrs, Casey. 2004. "Locally-Led Advance Mobile Aid." *Journal of Humanitarian Assistance,* November 15. http://sites.tufts.edu/jha/archives/824, accessed October 18, 2011.

Barrs, Casey. 2010. "Preparedness Support: Helping Brace Local Staff, Partners and Beneficiaries for Violence." Washington, DC: The Cuny Center. http://cunycenter .org/files/Preparedness%20Support7.pdf, accessed October 18, 2011.

Bazerman, Max H., and Michael Watkins. 2004. *Predictable Surprises: The Disasters You Should Have Seen Coming and How to Prevent Them.* Cambridge, MA: Harvard Business Press.

Beah, Ishmael. 2007. *A Long Way Gone: Memoirs of a Boy Soldier.* 1st ed. New York: Farrar, Straus and Giroux.

Berdal, M., and D. Malone, eds. 2000. *Greed and Grievance: Economic Agendas in Civil Wars.* Boulder, CO: Lynne Reinner.

Blénesi, Éva. 1998. "Ethnic Early Warning Systems and Conflict Prevention." Global Security Fellows Initiative, Occasional Paper No. 11. Cambridge, UK: University of Cambridge, Faculty of Social and Political Sciences.

Blomberg, S. Brock and Gregory D. Hess. 2002. "The Temporal Links between Conflict and Economic Activity." *Journal of Conflict Resolution* 46 (1): 74–90.

Bock, Joseph G. 1997. "Communal Conflict, NGOs, and the Power of Religious Symbols." *Development in Practice* 7 (1): 17–25.

Bock, Joseph G. 1999. The Harmony Project: Peace Building amid Poverty in India. In *Do No Harm: How Aid Can Support Peace—Or War,* ed. Mary Anderson, 119–130. Boulder, CO; London: Lynne Renner.

Bock, Joseph G. 2001a. *Sharpening Conflict Management: Religious Leadership and the Double-Edged Sword.* Westport, CT: Praeger.

Bock, Joseph G. 2001b. "Track II Anti-Incitement." *New Routes: A Journal of Peace Research and Action* 6 (1): 28–34.

Bock, Joseph G. 2005. "Review of the Moral Imagination: The Art and Soul of Building Peace, by John Paul Lederach." *Development in Practice* 15 (6) (November): 801–802.

Bock, Joseph G. 2009. "The Efficacy of Violence Mitigation: A Second Look Using Time-Series Analysis." *Political Geography* 28: 266–270.

Bock, Joseph G., and Mary B. Anderson. 1999. "Dynamite under the Intercommunal Bridge: How Can Aid Agencies Help Defuse It?" *Journal of Peace Research* 36 (3) (May 1): 325–338.

Bock, Joseph G., Patricia Lawrence, and Timmo Gaasbeek. 2009. Foundation for Co-Existence's Human Security Program in the Eastern Province: Final Narrative Report and Impact Assessment. In *Responding to Civil War: An Examination of a Third Generation Early Warning and Early Response System,* ed. Kumar Rupesinghe, 194–239. Colombo, Sri Lanka: The Foundation for Co-Existence.

Bock, Joseph G., and Clark McCauley. 2003. "A Call to Lateral Mission: Mobilizing Religious Authority Against Ethnic Violence." *Mission Studies* 22 (40): 9–34.

Bond, Doug, Joe Bond, Sean O'Brien, and Vladimir Petroff. 2004. *Forecasting Turmoil in Indonesia: An Application of Hidden Markov Models.* New York: Carnegie Corporation.

Bond, Doug, Joe Bond, Churl Oh, J. Craig Jenkins, and Charles Lewis Taylor. 2003. "Integrated Data for Events Analysis (IDEA): An Event Typology for Automated

Events Data Development." *Journal of Peace Research* 40 (6) (November 1): 733–745.

Bond, Doug, Craig Jenkins, Charles Taylor, and Kurt Schock. 1997. "Mapping Mass Political Conflict and Civil Society: Issues and Prospects for the Automated Development of Event Data." *Journal of Conflict Resolution* 41 (4): 553–579.

Bond, Doug, and Patrick Meier. 2006. "Resource Scarcity and Pastoral Armed Conflict in the Horn of Africa." Paper presented at the International Studies Association, Town and Country Resort and Convention Center, San Diego, CA, March 22.

Bowley, Graham. 2010. "Computers that Trade on the News." *The New York Times*, December 23, sec. B.

Bragg, Belinda, and Renat Shaykhutdinov. 2007. "Do Grievances Matter: An Experimental Examination of the Greed vs. Grievance Debate." Paper presented at the annual meeting of the International Society of Political Psychology, Classical Chinese Garden, Portland, OR, July 4.

Brass, Paul R. 1997. *Theft of an Idol: Text and Context in the Representation of Collective Violence*. Princeton, NJ: Princeton University Press.

Carment, David, and Karen Garner. 1998. "Early Warning and Conflict Prevention: Problems, Pitfalls and Avenues for Success." *Canadian Foreign Policy* 6 (2) (February): 103–118.

Carnegie Commission on Preventing Deadly Conflict. 1997. *Preventing Deadly Conflict*. New York: Carnegie Corporation of New York.

Center for International Media Assistance. 2007. *Community Radio: Its Impact and Challenges to its Development*. Washington, DC: National Endowment for Democracy. http://cima.ned.org/sites/default/files/CIMA-Community_Radio-Working_Group _Report_0.pdf, accessed October 17, 2011.

Cheetham, Robert, David Felcan, and Tony Smith. 2006. "Crime Density Change and Analysis: A Hypergeometric Approach." Unpublished manuscript. Philadelphia, PA: Avencia Incorporated.

Cheetham, Robert, and Sean McGinnis. 2009. *Hunchlab: Pierce County Crime Early Warning System*. Unpublished report. Philadelphia, PA: Avencia Incorporated.

Chenoweth, Erica. 2011. "Give Peaceful Resistance a Chance." *The New York Times*, March 9, A31.

Chenoweth, Erica, and Maria J. Stephan. 2011. *Why Civil Resistance Works: The Strategic Logic of Nonviolent Conflict*. New York: Columbia University Press.

Chicago Police Department. 2010. 2009 Annual Report. Chicago Police Department, Bureau of Administrative Services, Research and Development Division.

https://portal.chicagopolice.org/portal/page/portal/ClearPath/News/09AR.pdf, accessed October 24, 2011.

Chigas, Diana. 2006. "Has Peacebuilding Made a Difference in Kosovo? A Study of the Effectiveness of Peacebuilding in Preventing Violence: Lessons Learned from the March 2004 Riots in Kosovo." Cambridge, MA: CDA Collaborative Learning Projects.

Chowdurhy, Mridul. 2008. The Role of the Internet in Burma's Safron Revolution. Cambridge, MA: The Berkman Center for Internet and Society at Harvard University. http://cyber.law.harvard.edu/publications/2008/Role_of_the_Internet_in_Burmas _Saffron_Revolution, accessed October 25, 2011.

Cockell, John. 2003. Early Warning Analysis and Policy Planning in UN Preventive Action. In *Conflict Prevention: Path to Peace or Grand Illusion*, ed. David Carment and Albrecht Schnabel, 182–206. Blue Ridge Summit, PA: United Nations Publications.

Collier, Paul, Lisa Chauvet, and Haavard Hegre. 2008. The Security Challenges in Conflict-Prone Countries. Copenhagen Consensus Center, Copenhagen Consensus 2008 Challenge Paper. http://www.humansecuritygateway.com/documents/CP_Collier _securitychallengeinconflictpronecountries.pdf, accessed February 10, 2012.

Collier, Paul, and Anke Hoeffler. 2004. "Greed and Grievance in Civil War." *Oxford Economic Papers* 56 (4) 563–595.

Cross, Nigel. 2001. "Designerly Ways of Knowing: Design Discipline Versus Design Science." *Design Issues* 17 (3) (Summer): 49–52.

Currion, Paul. 2011. "Conclusion." In *Peacebuilding in the Information Age: Sifting Hype from Reality*, ed. Daniel Stouffacher, Barbara Weekes, Urs Gasser, Collin Maclay, and Michael Best, 39–42. Geneva: ICT4Peace Foundation in cooperation with the Berkman Center for Internet & Society at Harvard University and the Georgia Institute of Technology. January. http://ict4peace.org/wp-content/uploads/ 2011/01/Peacebuilding-in-the-Information-Age-Sifting-Hype-from-Reality.pdf, accessed October 21, 2011.

Darley, John M. 2000. Bystander Phenomena. In *Encyclopedia of Psychology*, vol. 1, ed. Alan Kazden, 493–495. Washington, DC: American Psychological Association.

Darley, John, and Bibb Latané. 1968. "Bystander Intervention in Emergencies: Diffusion of Responsibility." *Journal of Personality and Social Psychology* 8: 377–383.

Davies, John L. 1998. "The Global Event-Data System Coder's Manual." College Park, MD: Center for International Development and Conflict Management, Department of Government and Politics, University of Maryland. http://www -rohan.sdsu.edu/~alexseev/RussiaInAsia/GEDSCodebook800_3.pdf, accessed October 14, 2011.

Davies, John L., and Ted Robert Gurr, eds. 1998. *Preventive Measures: Building Risk Assessment and Crisis Early Warning Systems*. Lanham, MD: Rowman & Littlefield.

DeRouen Jr., Karl, and Shaun Goldfinch. 2005. "Putting the Numbers to Work: Implications for Violence Prevention." *Journal of Peace Research* 42 (1): 27–45.

Diamond, Larry. 2011. "A Fourth Wave or False Start? Democracy after the Arab Spring." *Foreign Affairs*, May 21. http://www.foreignaffairs.com/articles/67862/larry -diamond/a-fourth-wave-or-false-start, accessed July 11, 2011.

Earman, John. 1992. *Bayes Or Bust? A Critical Examination of Bayesian Confirmation Theory*. Cambridge, MA: MIT Press.

Easwaran, Eknath. 1984. *A Man to Match His Mountains: Badshah Khan, Nonviolent Soldier of Islam*. Petaluma, CA: Nilgiri Press.

European Commission's Directorate-General for Humanitarian Aid. 2004. *Report on Security for Humanitarian Personnel: Standards and Practices for the Security for Humanitarian Personnel and Advocacy for Humanitarian Space*. Brussels: European Commission's Humanitarian Aid Office. http://www.humanitarianinfo.org/srilanka/ infocentre/security/docs/echo2.pdf, accessed October 20, 2011.

Economist, The. 1993. "The Hindu Upsurge: The Road to Ayodhya." *The Economist*, February 6, 21–23.

Economist, The. 2006. "Mobiles, Protests and Pundits." *The Economist*, October 28, 71.

Economist, The. 2009. "Obituary: Velupillai Prabhakaran, commander of the Tamil Tigers, died on May 18th, aged 54." *The Economist*, May 21.

Economist, The. 2011. "Mobile Services in Poor Countries: Not Just Talk: Clever Services on Cheap Mobile Phones Make a Powerful Combination—Especially in Poor Countries." *The Economist*, January 27. http://www.economist.com/node/ 18008202?story_id=18008202, accessed April 12, 2011.

Ekine, Sokari. 2009. *SMS Uprising: Mobile Phone Activism in Africa*. Oxford, UK: Fahamu Books.

European Centre for Conflict Prevention. 1999. *People Building Peace: 35 Inspiring Stories from around the World*. Utrecht, Netherlands: European Centre for Conflict Prevention.

Festinger, Leon. 1957. *A Theory of Cognitive Dissonance*. Stanford, CA: Stanford University Press.

Fischer, Peter, Joachim I. Kruegerb, Tobias Greitemeyerc, Claudia Vogrincicd, Andreas Kastenmüllere, Dieter Freyf, Moritz Heened, Magdalena Wicherd, and Martina Kainbacherd. 2011. "The Bystander-Effect: A Meta-Analytic Review on Bystander

Intervention in Dangerous and Non-Dangerous Emergencies." *Psychological Bulletin* 137 (4): 517–537.

Fisher, Roger, and William Ury. 1981. *Getting to Yes*. Boston: Houghton Mifflin.

Flake, Gary W. 1998. *The Computational Beauty of Nature: Computer Explorations of Fractals, Chaos, Complex Systems, and Adaptation*. Cambridge, MA: MIT Press.

Ford, David N. 1999. "A Behavioral Approach to Feedback Loop Dominance Analysis." *System Dynamics Review* 15 (1): 3–36.

Freire, Paulo. 1997. *Pedagogy of the Oppressed*. Reprint. New York: Continuum.

García, Alejandro. 1996. *Hijos de La Violencia: Campesinos de Colombia Sobreviven a "Golpes" De Paz*. Barcelona: Libros de la Catarata.

George, A. L., and J. E. Holl. 1997. *The Warning Response Problem and Missed Opportunities in Preventive Diplomacy. A Report to the Carnegie Commission on Preventing Deadly Conflict*. New York: Carnegie Corporation of New York.

Gilligan, James. 1996. *Violence: Our Deadly Epidemic and Its Causes*. New York: G. P. Putnam's Sons.

Giridharadas, Anand. 2010. "Africa's Gift to Silicon Valley: How to Track a Crisis." *The New York Times*, March 14, WK3.

Goldstone, Jack A. 2008. "Pathways to State Failure." *Conflict Management and Peace Science* 25 (4): 285–296.

Goldstone, Jack A. 2009. "Rethinking Revolutions: Integrating Origins, Processes, and Outcomes." *Comparative Studies of South Asia, Africa and the Middle East* 29 (1): 8–32.

Goldstone, Jack A., et al. 2010. "A Global Model for Forecasting Political Instability." *American Journal of Political Science* 54 (1) (January): 190–208.

Gopin, Marc. 2000. *Between Eden and Armageddon: The Future of World Religions, Violence and Peacemaking*. Oxford, UK: Oxford University Press.

Gowing, Nik. 2009. *Skyful of Lies and Black Swans: The New Tyranny of Shifting Information Power in Crises*. Oxford, UK: Oxford University Press.

Guarino, Mark. 2010. "Behind Chicago's High-Crime Summer: Persistent Street Gang Violence." *The Christian Science Monitor*, August 30. http://www.csmonitor.com/USA/Justice/2010/0830/Behind-Chicago-s-high-crime-summer-persistent-street-gang-violence, accessed October 22, 2011.

Gujer, Eric. 2011. "Intelligence of the Masses or Stupidity of the Herd?" In *Peacebuilding in the Information Age: Sifting Hype from Reality*, ed. Daniel Stouffacher, Barbara Weekes, Urs Gasser, Collin Maclay, and Michael Best, 23–25. Geneva: ICT4Peace

Foundation in cooperation with the Berkman Center for Internet & Society at Harvard University and the Georgia Institute of Technology. January. http://ict4peace.org/wp-content/uploads/2011/01/Peacebuilding-in-the-Information-Age-Sifting-Hype-from-Reality.pdf, accessed October 21, 2011.

Gurr, T. R. 1970. *Why Men Rebel*. Princeton, NJ: Princeton University Press.

Gurr, T., and B. Harff. 1996. *Early Warning of Communal Conflict and Genocide*. Tokyo: United Nations University Press.

Gurr, Ted Robert, and Barbara Harff. 1998. "Systematic Early Warning of Humanitarian Emergencies." *Journal of Peace Research* 35 (5): 551–579.

Hardin, Russell. *Collective Action*. 1982. Baltimore, MD: Johns Hopkins University Press.

Harvard Humanitarian Initiative. 2011. *Disaster Relief 2.0: The Future of Information Sharing in Humanitarian Emergencies*. Washington, DC, and Berkshire, UK: UN Foundation & Vodafone Foundation Technology Partnership. http://www.unocha.org/top-stories/all-stories/disaster-relief-20-future-information-sharing-humanitarian-emergencies, accessed October 22, 2011.

Hatchuel, Armand. 2002. "Towards a Design Theory and Expandable Rationality: The Unfinished Program of Herbert Simon." *Journal of Management and Governance* 5 (3–4): 1–12.

Hattotuwa, Sanjana, and Daniel Stauffacher. 2010. "Cross-fertilisation of UN Common Operational Datasets and Crisismapping." Geneva: ICT4Peace Foundation. October. http://ict4peace.org/wp-content/uploads/2010/10/UN-and-CrisisMapping.pdf, accessed October 22, 2011.

Hauben, Michael, and Ronda Hauben. 1997. *Netizens: On the History and Impact of Usenet and the Internet*. Washington, DC: Institute of Electrical and Electronic Engineers Computer Society Press.

Heinzelman, Jessica, and Carol Waters. 2010. *Crowdsourcing Crisis Information in Disaster-Affected Haiti*. Washington, DC: United States Institute of Peace.

Herrmann, Roy. 2003. "Mid-Term Review of a Canadian Security Deployment to the UNHCR Programme in Guinea." Geneva: United Nations High Commissioner for Refugees.

Himelfarb, Sheldon, with Cecilia Paradi-Guilford. 2010. "Can You Help Me Now? Mobile Phones and Peacebuilding in Afghanistan." United States Institute of Peace Special Report, Washington, DC. http://www.usip.org/files/resources/SR%20259%20-%20Can%20You%20Help%20Me%20Now.pdf, accessed July 11, 2011.

Hobbs, Jerry R., Douglas Appelt, John Bear, David Israel, Megumi Kameyama, Mark Stickel, and Mabry Tyson. 1992. FASTUS: Extracting Information from Natural-

Language Texts. In *Finite-State Language Processing*, ed. Emmanuel Roche and Yves Schabes, 383–406. Cambridge, MA: MIT Press.

Hoffman, Benjamin, and Evan Hoffman. 2006. *Preventing Political Violence: Towards a Model for Catalytic Action*. Ottawa, Ontario: Canadian International Institute of Applied Negotiation.

Holland, R., R. M. Meertens, and M. Van Vugt. 2002. "Dissonance on the Road: Self-Esteem as a Moderator of Internal and External Oriented Modes of Self-Justification." *Personality and Social Psychology Bulletin* 12: 1713–1724.

Horowitz, Donald L. 2001. *The Deadly Ethnic Riot*. Berkeley: University of California Press.

Howard, Philip N. 2011. *The Digital Origins of Dictatorship and Democracy: Information Technology and Political Islam*. Oxford Studies in Digital Politics. Oxford, UK: Oxford University Press.

ICT4Peace Foundation. 2011. "Managing the Accelerating Complexity of Humanitarian Response. In *Peacebuilding in the Information Age: Sifting Hype from Reality*, ed. Daniel Stouffacher, Barbara Weekes, Urs Gasser, Collin Maclay, and Michael Best, 31–33. Geneva: ICT4 Peace Foundation in cooperation with the Berkman Center for Internet & Society at Harvard University and the Georgia Institute of Technology. January. http://ict4peace.org/wp-content/uploads/2011/01/Peacebuilding-in-the-Information-Age-Sifting-Hype-from-Reality.pdf, accessed October 21, 2011.

International Federation of Red Cross and Red Crescent Societies. 2009. *World Disasters Report 2009—Focus on Early Warning, Early Action*. Geneva: International Federation of Red Cross and Red Crescent Societies.

IRIN Humanitarian News and Analysis. 2010. "Kenya: SOS by SMS." Humanitarian News and Analysis. New York: United Nations, August 3. http://www.irinnews.org/Report.aspx?REportID=90050, accessed October 20, 2011.

Johansen, Robert. 1997. "Radical Islam and Nonviolence: A Case Study of Religious Empowerment and Constraint among Pushtuns." *Journal of Peace Research* 34 (1): 53–71.

Justino, Patricia. 2007. "Carrot or Stick? Redistributive Transfers versus Policing in Contexts of Civil Unrest." Brighton, UK: Households in Conflict Network. http://ideas.repec.org/p/hic/wpaper/33.html, accessed February 12, 2012.

Kakar, Sudhir. 2001. "Political Science: Killing Other People." *Science* 291 (5511): 2097.

King, Gary, and Will Lowe. 2003. "An Automated Information Extraction Tool for International Conflict Data with Performance as Good as Human Coders: A Rare Events Evaluation Design." *International Organization* 57 (3): 617–642.

Korenblum, Jacob, and Bieta Andemariam. 2011. Cell Phones and Conflict Zones: How Souktel Uses SMS Technology to Empower and Aid in Conflict-Affected Communities. In *Mobile Technologies for Conflict Management: Online Dispute Resolution, Governance, Participation*, ed. Marta Poblet, 67–80. New York: Springer.

Korf, Benedikt. 2005. "Rethinking the Greed-Grievance Nexus: Property Rights and the Political Economy of War in Sri Lanka." *Journal of Peace Research* 42 (2) (March 1): 201–217.

Kreps, Sarah E. 2010. Social Networks and Technology in the Prevention of Crimes against Humanity. In *Mass Atrocity Crimes: Preventing Future Outrages*, ed. Robert I. Rotberg, 175–191. Washington, DC: Brookings Institution Press.

Kurtz, C., and D. Snowden. 2003. "The New Dynamics of Strategy: Sense Making in a Complex Complicated World." *IBM Systems Journal* 42 (3): 462–483.

Lacey, Marc. 2005. "Beyond the Bullets and Blades." *The New York Times*, March 20.

Land, Molly Beutz. 2009. "Networked Activism." *Harvard Human Rights Journal* 22: 205–243.

Leaning, Jennifer. 2010. The Use of Patterns in Crisis Mapping to Combat Mass Atrocity Crimes. In *Mass Atrocity Crimes: Preventing Future Outrages*, ed. Robert I. Rotberg, 192–219. Washington, DC: Brookings Institution Press.

Lederach, John Paul. 1997. *Building Peace: Sustainable Reconciliation in Divided Societies*. Washington, DC: United States Institute of Peace.

Lederach, John Paul. 2005. *The Moral Imagination: The Art and Soul of Building Peace*. New York: Oxford University Press.

Lederach, John Paul, and R. Scott Appleby. 2010. Strategic Peacebuilding: An Overview. In *Strategies of Peace: Transforming Conflict in a Violent World*, ed. Daniel Philpott and Gerard F. Powers, 19–44. New York: Oxford University Press.

Lipsey, R. G., and Kelvin Lancaster. 1956. "The General Theory of Second Best." *Review of Economic Studies* 24 (1): 11–32.

Logan, B. I., and W. G. Moseley. 2001. "Conceptualizing Hunger Dynamics: A Critical Examination of Two Famine Early Warning Methodologies in Zimbabwe." *Applied Geography* 21 (3) (July): 223–248.

Lund, Michael S. 1996. *Preventing Violent Conflicts: A Strategy for Preventive Diplomacy*. Washington, DC: United States Institute of Peace Press.

Lund, M. 1998. *Not Only When to Act, But How: From Early Warning to Conflict Prevention: Path to Peace or Grand Illusion?* New York: United Nations.

Lyke, Susan A., and Joseph G. Bock. 1995. "Orangi Pilot Project—Research and Training Institute of Karachi, Pakistan." Case Study No. 9, Local Capacities for Peace Project. Cambridge, MA: Collaborative for Development Action. http://www.cdainc

.com/cdawww/pdf/casestudy/dnh_pakistan_case_study_set_1_Pdf.pdf, accessed October 18, 2011.

Maasen, Kristel. 2007. *Mobilising Early Response to Prevent Violent Conflict: An Overview of Obstacles to Early Response*. The Hague, Netherlands: Global Partnership for the Prevention of Armed Conflict.

Mandara, Christina Geoffrey. 2007. "Participatory GIS in Mapping Local Context of Conflicts over Pastoral Resources: A Case Study of Duru Haitemba–Babati, Tanzania." Thesis. Enschede, Netherlands. http://www.itc.nl/library/papers_2007/msc/nrm/mandara.pdf, accessed October 23, 2011.

Mathews, Dylan. 2001. *War Prevention Works: 50 Stories of People Resolving Conflict*. Oxford, UK: Oxford Research Group.

Matveeva, Anna. 2006. "Early Warning and Early Response: Conceptual and Empirical Dilemmas." Issue Paper 1. The Hague, Netherlands: European Centre for Conflict Prevention/International Secretariat of the Global Partnership for the Prevention of Armed Conflict. September.

McCauley, Clark. 2000. "How President Bush Moved the U.S. into the Gulf War: Three Theories of Group Conflict and the Construction of Moral Violation." *Journal for the Study of Peace and Conflict* 3 (1): 32–42.

McCauley, Clark, and Joseph G. Bock. 2004. Why Does Violence Trump Peace Building? Negativity Bias in Intergroup Relations. In *The Psychology of Ethnic and Cultural Conflict*, ed. Yueh-Ting Lee, Clark McCauley, Fathali Moghaddam, and Stephen Worchel, 273–288. Westport, CT: Praeger.

McClelland, Charles. 1972. *International Events Interaction Analysis: Some Research Considerations*. Beverly Hills, CA: Sage Publications.

McCombs, M. E., and D. L. Shaw. 1972. "The Agenda-Setting Function of Mass Media." *Public Opinion Quarterly* 36: 176–187.

Meadows, Donella H., and Jennifer M. Robinson. 2002. "The Electronic Oracle: Computer Models and Social Decisions." *System Dynamics Review* 18: 271–308.

Meier, Patrick. 2008. "A Conversation on Early Warning with Howard Adelman." http://earlywarning.wordpress.com/2008/11/08/a-conversation-on-early-warning-with-howard-adelman, accessed October 17, 2011.

Meier, Patrick. 2011. "Early Warning Systems and the Prevention of Violent Conflict." In *Peacebuilding in the Information Age: Sifting Hype from Reality*, ed. Daniel Stouffacher, Barbara Weekes, Urs Gasser, Collin Maclay, and Michael Best, 12–15. Geneva: ICT4 Peace Foundation in cooperation with the Berkman Center for Internet & Society at Harvard University and the Georgia Institute of Technology. January. http://ict4peace.org/wp-content/uploads/2011/01/Peacebuilding-in-the-Information-Age-Sifting-Hype-from-Reality.pdf, accessed October 21, 2011.

Meier, Patrick, Doug Bond, and Joe Bond. 2007. "Environmental Influences on Pastoral Conflict in the Horn of Africa." *Political Geography* 26: 716–735.

Meier, Patrick, and Jennifer Leaning. 2009. "Applying Technology to Crisis Mapping and Early Warning in Humanitarian Settings." Working Paper Series. Cambridge, MA: Harvard Humanitarian Initiative. September. http://hhi.harvard.edu/images/resources/hhiworkingpapermeierleaning.pdf, accessed October 22, 2011.

Meng, Jude Chua Soo. 2009. "Donald Schön, Herbert Simon and The Sciences of the Artificial." *Design Studies* 30 (1) (January): 60–68.

Merton, R. K. 1938. "Social Structure and Anomie." *American Sociological Review* 3 (5): 672–682.

Morozov, Evgeny V. 2011. *The Net Delusion: The Dark Side of Internet Freedom*. New York: Public Affairs.

Muhumuza, Joseph, and Dennis T. Bataringaya. 2009. "Mapping of Civil Society Organizations (CSOs) Involved in Conflict Prevention, Management and Resolution (CPMR) Work in the Ugandan Side of the Karamoja Cluster." Unpublished report submitted to CEWARN/IGAD, Addis Ababa, Ethiopia, http://www.cewarn.org/index.php?option=com_docman&task=doc_download&gid=18&Itemid=87, accessed October 17, 2011.

Murshed, Syed Mansoob. 2008. "Indicators of Potential Conflict." MICROCON Policy Briefing No. 1. Brighton, UK: Institute of Development Studies at the University of Sussex.

Murshed, Syed Mansoob, and Mohammad Zulfan Tadjoeddin. 2007. "Reappraising the Greed and Grievance Explanations for Violent Internal Conflict." MICROCON Research Working Paper No. 2. Brighton, UK: Institute of Development Studies at the University of Sussex. September.

National Gang Intelligence Center. 2009. *National Gang Threat Assessment 2009*. Johnstown, PA: National Drug Intelligence Center.

National Institute of Urban Affairs. 1994. *Urban Environmental Maps for Bombay, Delhi, Ahmedabad, Vadodara*. New Delhi, India: National Institute of Urban Affairs.

Nyheim, David. 2008. "Can Violence, War and State Collapse Be Prevented? The Future of Operational Conflict Early Warning and Response Systems." Presented at the 10th Meeting of the DAC Fragile States Group and Conflict, Peace and Development Co-Operation. Paris: International Energy Agency Headquarters.

Nyheim, David. 2009. Three Generations in Early Warning: Challenges and Future Directions. In *Responding to Civil War: An Examination of a Third Generation Early Warning and Early Response System*, ed. Kumar Rupesinghe, 38–59. Colombo, Sri Lanka: The Foundation for Co-Existence.

O'Brien, Sean. 2002. "Anticipating the Good, the Bad, and the Ugly: An Early Warning Approach to Conflict and Instability Analysis." *Journal of Conflict Resolution* 46 (6): 791–811.

O'Brien, Sean P. 2010. "Crisis Early Warning and Decision Support: Contemporary Approaches and Thoughts on Future Research." *International Studies Review* 12 (1) (March 9): 87–104. http://onlinelibrary.wiley.com/doi/10.1111/j.1468-2486.2009 .00914.x/full, accessed October 25, 2011.

O'Callaghan, Sorcha, and Sara Pantuliano. 1999. *Protection of Internally Displaced Persons.* New York: Inter-Agency Standing Committee Policy Paper. December.

O'Callaghan, Sorcha, and Sara Pantuliano. 2007. Protective Action: Incorporating Civilian Protection into Humanitarian Response. HPG Report No. 26, Humanitarian Policy Group, Overseas Development Institute, London, December. http://www.odi .org.uk/resources/download/1020.pdf, accessed October 23, 2011.

Olson, Mancur. 1971. *The Logic of Collective Action: Public Goods and the Theory of Groups.* Cambridge, MA: Harvard University Press.

Orwell, George. 1949. *1984, a Novel.* 1st American ed. New York: Harcourt, Brace. In the Herman Finkelstein Collection. Library of Congress, Washington, DC.

Patel, Pravin. 1995. Communal Riots in Contemporary India: Towards a Sociological Explanation. In *Crisis and Change in Contemporary India,* ed. Upendra Baxi and Bhikha Parekh, 370–399. New Delhi: Sage.

Payson Conflict Study Group. 2001. *A Glossary on Violent Conflict: Terms and Concepts Used in Conflict Prevention, Mitigation, and Resolution in the Context of Disaster Relief and Sustainable Development.* New Orleans: Payson Center for International Development and Technology Transfer, Tulane University.

Philadelphia Police Department. n.d. "Philadelphia Police Spike Detector (Official Manual)." Unpublished manuscript. Philadelphia, PA.

Poblet, Marta, ed. 2011. *Mobile Technologies for Conflict Management: Online Dispute Resolution, Governance, Participation.* New York: Springer.

Prakash, Cedric. 1994a "Communal Violence and Its Impact on the Urban Poor." Paper presented at the seminar Communal Violence and Communalism in Western India, organized by the Centre for Social Studies, Surat, India.

Prakash, Cedric. 1994b. *Annual Report.* Unpublished. Ahmedabad, India: St. Xavier's Social Services Society.

Rapoport, David C. 1993. Comparing Militant Fundamentalist Movements and Groups. In *Fundamentalisms and the State: Remaking Politics, Economics, and Militance,* ed. Martin E. Marty and R. Scott Appleby, 429–461. Chicago: University of Chicago Press.

Rudolph, Susanne Hoeber, and Lloyd I. Rudolph. 1993. "Modern Hate." *New Republic* 208 (12) (March 22): 24–29.

Rupesinghe, Kumar, and Michiko Kuroda, eds. 1992. *Early Warning and Conflict Resolution.* New York: St. Martin's Press.

Salazar, Oscar, and Jorge Soto. 2011. How to Crowdsource Election Monitoring in 30 Days: The Mexican Experience. In *Mobile Technologies for Conflict Management: Online Dispute Resolution, Governance, Participation,* ed. Marta Poblet, 55–66. New York: Springer.

Schmeidl, Susanne. 1998. "The Early Warning of Humanitarian Disasters: Problems in Building an Early Warning System." *International Migration Review* 32 (2): 471–486.

Schmeidl, Susanne. 2001. "Early Warning and Integrated Response Development." *Romanian Journal of Political Science* 2: 4–50.

Schmeidl, S., and J. C. Jenkins. 1998. "The Early Warning of Humanitarian Disasters: Problems in Building an Early Warning System." *International Migration Review* 32: 471–486.

Schmeidl, Susanne, and Cirû Mwaûra, eds. 2002. *Early Warning and Conflict Management in the Horn of Africa.* 1st ed. Lawrenceville, NJ: The Red Sea Press.

Schmeidl, Susanne, and Eugenia Piza-Lopez. 2002. *Gender and Conflict Early Warning: A Framework for Action.* London: International Alert.

Schön, Donald A. 1983. *The Reflective Practitioner: How Professionals Think in Action.* New York: Basic Books.

Schrodt, Philip A. 2000. Pattern Recognition of International Crises Using Hidden Markov Models. In *Political Complexity: Nonlinear Models of Politics,* ed. Diana Richards, 296–328. Ann Arbor: University of Michigan Press.

Schrodt, Philip A., and Deborah Gerner. 1994."Validity Assessment of a Machine-Coded Event Data Set for the Middle East, 1982–92." *American Journal of Political Science* 38 (3): 825–854.

Schrodt, Philip A., and Deborah J. Gerner. 2000. "Cluster-Based Early Warning Indicators for Political Change in the Contemporary Levant." *American Political Science Review* 94 (4) (December): 803–817.

Schwartz, Peter. 1996. *The Art of the Long View: Planning for the Future in an Uncertain World.* New York: Doubleday.

Senge, Peter M. 2006. *The Fifth Discipline: The Art and Practice of the Learning Organization.* New York: Doubleday/Currency.

Sharp, Gene. 2003. *From Dictatorship to Democracy: A Conceptual Framework for Liberation.* 2nd ed. Boston: The Albert Einstein Institution.

Shirky, Clay. 2008. *Here Comes Everybody: The Power of Organizing without Organizations*. New York: Penguin Press.

Simon, Herbert, and Donald Schon. 1996. *The Sciences of the Artificial*. Cambridge, MA: MIT Press.

Siror, Joseph K., Sheng Huanye, Dong Wang, and Wu Jie. 2009. "Use of RFID Technologies to Combat Cattle Rustling in the East Africa." Fifth International Joint Conference on INC, IMS and IDC, Seoul, Korea, August 25–27. http://www.computer.org/portal/web/csdl/doi/10.1109/NCM.2009.146, accessed October 17, 2011.

Skogan, Wesley G., Susan M. Hartnett, Natalie Bump, and Jill Dubois. 2008. With the assistance of Ryan Hollon and Danielle Morris. *Evaluation of CeaseFire-Chicago*. Northwestern University, Institute for Policy Research: National Institute of Justice, Office of Justice Programs. http://www.northwestern.edu/ipr/publications/cease-fire_papers/mainreport.pdf, accessed February 22, 2010.

Snyder, Jack L. 2000. *From Voting to Violence: Democratization and Nationalist Conflict*. New York: W. W. Norton.

Steele, David. 1998. Conflict Resolution among Religious People in Bosnia and Croatia. In *Religion and the War in Bosnia*, ed. Paul Mojzes, 246–253. Atlanta, GA: Scholars Press.

Steele, David. 1999. Ecumenical Community Building and Conflict Resolution Training in the Balkans. In *Training to Promote Conflict Management: USIP-Assisted Training Projects*, ed. David Smock, 22–27. Washington, DC: U.S. Institute of Peace.

Stewart, Frances. 2009. "Religion versus Ethnicity as a Source of Mobilisation: Are There Differences?" MICROCON Research Working Paper No. 18. Brighton, UK: Institute of Development Studies at the University of Sussex. October.

Sunstein, Cass R. 2006. *Infotopia: How Many Minds Produce Knowledge*. New York: Oxford University Press.

Surowiecki, James. 2004. *The Wisdom of the Crowds: Why the Many Are Smarter Than the Few and How Collective Wisdom Shapes Business, Economies, Societies, and Nations*. New York: Anchor Books.

Tenenbaum, Joshua B. 1999. Baysean Modeling of Human Concept Learning. In *Advances in Neural Information Processing Systems 11*, ed. Michael S. Kearns, Sara A. Solla, and David A. Cohn, 59–68. Cambridge, MA: MIT Press.

Theodore, Jesse. 2009. "Predictive Modeling Becomes a Crime-Fighting Asset." *Law Officer Journal* 5 (2): 1.

Thomson, M. 2006. "Malaria Early Warnings Based on Seasonal Climate Forecasts from Multi-Model Ensembles." *Nature* 439: 576–579.

Toft, Monica Duffy. 2003. *The Geography of Ethnic Violence: Identity, Interests, and the Indivisibility of Territory.* Princeton, NJ: Princeton University Press.

Varshney, Ashutosh. 2001. "Ethnic Conflict and Civil Society: India and Beyond." *World Politics* 53 (April): 362–398.

Weidmann, Nils B., and Doreen Kuse. 2009. "WarViews: Visualizing and Animating Geographic Data on Civil War." *International Studies Perspectives* 10 (1) (January 29): 36–48.

Weinberg, Tamar. 2009. *The New Community Rules: Marketing on the Social Web.* Sebastopol, CA: O'Reilly Media.

Wertman, Stephen. 2009. "Bleeding for Humanity—Humanitarian Intervention Is Politics: A New Doctrine." *Harvard International Review.* June 17. http://hir.harvard.edu/bleeding-for-humanity, accessed October 18, 2011.

Wigger, Andreas. 2011. "Protection of Civilian Populations in Conflicts and Other Situations of Violence: The Challenges of Using ICTs." In *Peacebuilding in the Information Age: Sifting Hype from Reality,* ed. Daniel Stouffacher, Barbara Weekes, Urs Gasser, Collin Maclay, and Michael Best, 16–19. Geneva: ICT4 Peace Foundation in cooperation with the Berkman Center for Internet & Society at Harvard University and the Georgia Institute of Technology. January. http://ict4peace.org/wp-content/uploads/2011/01/Peacebuilding-in-the-Information-Age-Sifting-Hype-from-Reality.pdf, accessed October 21, 2011.

Wisner, Ben. 2006. *Let Our Children Teach Us: A Review of the Role of Education and Knowledge in Disaster Risk Reduction.* Bangalore: Books for Change.

Wohlgelernter, Elli, and David Rudge. 1997. "Chief Rabbi Condemns Pig Plot in Letter to Arafat." *The Jerusalem Post,* December 29, 5.

Women's Commission for Refugee Women and Children. 2006. *Displaced Women and Girls at Risk: Risk Factors, Protection Solutions and Resource Tools.* New York: Women's Commission for Refugee Women and Children.

Woocher, Lawrence. 2009. *Preventing Violent Conflict: Assessing Progress, Meeting Challenges.* Washington, DC: United States Institute of Peace.

World Bank. 2006. *Engaging with Fragile States.* Washington, DC: World Bank.

World Bank. 2011. *World Development Report 2011: Conflict, Security, and Development.* Washington, DC: World Bank.

Wulf, Herbert, and Tobias Debiel. 2009. "Conflict Early Warning and Response Mechanisms: Tools for Enhancing the Effectiveness of Regional Organizations: A Comparative Study of the AU, ECOWAS, IGAD, ASEAN/ARF and PIF." Working paper no. 49—Regional and Global Axes of Conflict. Crisis States Working Papers Series

No. 2. London: London School of Economics. http://www.wulf-herbert.de/WP49 .pdf, accessed October 17, 2011.

Yiu, Chris, and Nick Mabey. 2005. *Countries at Risk of Instability: Practical Risk Assessment, Early Warning and Knowledge Management.* London: Prime Minister's Strategy Unit.

Zadeh, L. A. 1965. "Fuzzy Sets." *Information and Control* 8 (3): 338–353.

Zarrella, Dan. 2009. *The Social Media Marketing Book.* Sebastopol, CA: O'Reilly Media.

Index

Abebe, Betty, 6
Absorptive capacity, 168
Accelerators of conflict, 25
Ackleson, Jason, 11
Actionable information, 18, 37, 45, 106, 143, 200
Addis Ababa, Ethiopia, 3, 127–128
Adelman, Howard, 11, 127
Afghanistan, 108, 150, 193–194
Africa, 3, 32, 34, 42, 85, 105, 108, 115, 127–129, 131, 133, 140, 182, 190–191, 207
African Union (AU), 119
Agenda-setting function, 73
Agent-based early warning, 51, 53, 149
Aggregation of data, 136, 141
Ahimsa, 8
Ahmedabad, India, 12, 24, 57–61, 63–65, 67, 69, 71, 73, 75, 77–79, 82, 95, 155
Al Aqsa Mosque, 148–149
Al-Mi'rage, 148
Albanians, 28
Algorithms, 39, 48, 142, 192
Amara, Iraq, 172
Amman, Colonel Karuna, 93
Amman, Jordan, 93
Ampara, Sri Lanka, 93, 96–97, 140, 184–185
Amsterdam, The Netherlands, 115
Anderson, Mary B., 150

Appleby, R. Scott, 7, 30, 171
Arab Spring, 2, 7, 205
Arafat, Yasser, 149
Arbitration, 194
Architecture of early warning systems, 28, 35, 46, 52, 168
Areas of responsibility (AORs), 185, 188
Ariely, Dan, 111
Arizona, 85
Armed Revolutionary Front of Colombia (FARC), 158
Art used in peacebuilding, 64–65, 70–71, 79
Assam, India, 22
Association of Peasant Workers of Carrara (ATCC), 159
Atallah, Nabil, 149
Austin, Alexander, 183
Authority, use of, 3, 13, 33, 49, 52, 60, 75, 78, 127, 133, 147, 149, 169, 207
Automation, 12, 37, 42–43, 46, 52, 112, 137–138, 207
Avaaz, human rights web-based platform, 117
Avatars, 81–82, 189
Ayers, Michael D., 118
Ayodhya, India, 59, 72, 74
Azavea, 196

Badal code, 151–152, 159
Baidoa, Somalia, 181

Bakshi-Doron, Sephardi Chief Rabbi
Eliahu, 149
Balkans, 30, 32, 207
Ballpoint pen, as for picking a lock,
111
Bangkok, Thailand, 91
Bangladesh, 6, 58
Barrs, Casey, 156, 162–166, 168, 170,
173–174, 179
Bataringaya, Dennis T., 129
Batticaloa, Sri Lanka, 93, 96–98, 185
Bayesian modeling, 110
Bazerman, Max H., 37
Beah, Ishmael, 174
Berdal, M., 24
Bern, Switzerland, 40, 94, 102
Bharatiya Janata Party (BJP), 59
Bible, 26
Blénesi, Éva, 17, 147
Bock, Joseph G., 17, 25, 31, 46, 96, 97,
133,142,148,149,154,185
Bond, Doug, 40–41, 45, 50, 96, 98,
133, 137, 182
Bond, Joe, 40–41, 45, 51, 98, 133
Bootlegging, 26, 82
Bosnia, 11, 179
Botswana, 133
Bounded crowd feeding, 118,122,
124–126, 198, 206. See also Restricted
feeding
Bounded crowdsourcing, 113, 118, 122,
124–126, 198, 205–206. See also
Crowdsourcing
Bowley, Graham, 191–192
Bragg, Belinda, 24
Brass, Paul, 148
Brazil, 85
Brighton, England, 11
Broadband Internet access, 89
Brochures, 49, 71, 118
Brusset, Bruce, 183
Buddhist, 27, 92, 94, 149–150,
155, 185

Bump, Natalie, 85
Burma, 170–171, 174
Burundi, 41
Bystanders, as in witnessing violence,
31, 198

Calibration focus groups, 131
Cambodia, 11
Cambridge, Massachusetts, 119
Canada, 184
Caritas Neerlandica, 60
Caritas Sverige, Sweden, 60
Carlin, George, 91
Carment, David, 5
Carnegie Commission on Preventing
Deadly Conflict, 11, 28, 179
Cartographic Modeling Laboratory, 196
Caste, 65–66, 109
Categories of data, 4, 8, 18, 34–35,
41–42, 44–47, 52, 63, 92, 94–96,
98–99, 102, 109–110, 116, 120, 128,
135, 137, 139, 142, 177, 179, 186,
189–191, 193–194
Catholic Relief Services (CRS), 37, 60,
163, 166, 181
Cattle rustling, 13, 132–134, 184, 205.
See also Raids
Ceasefire, Chicago, 32–33, 35, 81–90
Cebemo, 60
Center for the Advancement of
Distance Education (CADE), 82, 89
Chalmers, M., 183
Chauvet, Lisa, 8, 10
Chechnya, 162
Cheetham, Robert, 196
Chenab River, Pakistan, 163
Chenoweth, Erica, 123
Chicago,Illinois 6, 12, 32, 81–87, 89–90
Chigas, Diana, 28
Civil society organizations (CSOs),
129–130, 133
Clan identity, 152–153
Class identity, 24, 31, 59, 72, 74, 151

Clergy, 26, 33, 86

Clickatell, 113

Clooney, George, 5

Cockell, John, 5

Coding of events data, 44–46, 92, 138

Cognitive dissonance theory, 25,
30–32, 35–36, 78, 204. *See also*
Moral dissonance

Collective goods dilemma, 197

Collective intelligence, 55, 106, 206

Collector, as in government official, 59,
71–72

Collier, Paul, 8, 10, 24

Colombia, 157–159

Colombo, Sri Lanka, 41, 94, 97, 99,
102, 155

Colonialism, 58, 152

Communal harmony, 63–65, 69–70,
72–73, 75–79

Communalism, 63–64, 69–70, 72–79,
109, 130, 151, 184

Community-based organizations
(CBOs), 6, 57, 94, 102, 118, 138, 181,
184, 187, 189, 206

Community organizing, 13, 60, 74, 78,
90, 205

Complex emergencies, 172, 185, 188

Computer application ("app"), 189,
194

Conflict early response, 111, 125

Conflict management, 11, 89, 91

Conflict prevention, management and
resolution (CPMR), 129

Conflict resolution, 18

Conflict transformation, 3, 111

Congo, Democratic Republic of (DRC),
5, 32, 85, 108, 162, 164, 180

Consensus building process, as toward
violence, 12, 15, 17–18, 20, 22, 24,
33–34, 36, 46, 79, 96, 98, 116, 124,
158, 204

Continental Early Warning System
(CEWS), African Union's, 190

Contractors, 129, 167

Coordinator for Reconstruction and
Stabilization, office of the U.S.
government, 183

Coreligionist communication in
anti-incitement, 28–32, 149, 159.
See also Intrafaith relations

Crimson Hexagon, 192

Croats, 32

Cross, Nigel, 35

Crowd feeding, 116, 118, 124, 206

Crowdsourcing, 4, 12–13, 48, 90,
105–126, 136, 141, 143, 184, 187,
197–199, 201, 203, 205–206. *See also*
Bounded crowdsourcing

Crowdsourcing honesty, 111, 125

Crowdsourcing the filter, 112, 125

Cuny, Fred, 162

Curating of data, 120

Currion, Paul, 23, 55, 91, 120, 161,
177, 179, 203

Cyprus, 167–168

Dallaire, Romeo, 186

Darfur, Sudan, 5, 164

Darley, John, 197

Data harvesting, 108

Data mining, 142, 203

Davies, John L., 11

Debiel, Tobias, 130, 142

De-escalation of conflict, 97, 204, 206

Defense, French government's
Secretariat General de la, 183

Democratic Republic of Congo
(DRC), 108

Democratization, 114, 119. *See also*
Netizen democracy

Demonstrations, as used in
nonviolence, 128, 166, 198

Denmark, 186, 188

Depletion of goodwill, 199

DeRouen Jr., Karl, 11

Diffusion of responsibility, 197–198

Digital straw, 48, 112, 114
DigitalGlobe, 5
Dillinger, John, 82
Diplomacy, 5, 15, 23, 94, 102, 147, 184
Disappearances in Sri Lanka, 93
Disasters, 2, 37, 39, 107–108, 119–120,
    122, 126, 161–163, 170, 182, 187,
    199, 205–206
Dissemination of information, 29, 109,
    115, 117, 119, 138, 140, 144, 156,
    206
Disutility, point of, 144–146
Divergence, as of moving averages, 50,
    131
Diversity, celebration of, 63, 77–78,
    141. See also Tolerance
Djibouti, 13, 127, 129
Domain analysis, 42–43
Donors and donations, 60, 92, 129,
    136, 140, 169, 182, 207

Easwaran, Eknath, 152
Economist, The, 8
Effectiveness of programs, 12, 22, 31,
    90, 99, 101, 131–132, 161, 163, 178,
    183–184, 190, 200
Efficiencies of programs, 138, 177,
    196, 206
Egypt, 1, 111, 122–124, 198, 205
Ekine, Sokari, 107
El Salvador, 11, 164
Elders, 94, 153, 194, 206
Elections, 13, 19–20, 105, 107–111,
    113, 115–117, 119, 121, 123–125,
    159, 169, 190, 205
Emergencies, 60–61, 85, 90, 108, 120,
    122, 144, 163, 166, 171–172, 185,
    188
Endaragalle, Dinidu, 6
England, 85
Enough Project, 5
Epidemics, 32, 62, 81
Epidemiologists, 105

Equity theory, 25
Eritrea, 127
Escalation, of conflict, 2, 8–13, 19, 25,
    30, 85, 90, 124, 147–148, 166, 180,
    184, 187, 198, 205
Ethics, 101, 153, 164
Ethiopia, 3, 6, 13, 127, 129–130
Ethnic cleansing, 32
Eubank, Dave, 171
Eurasia, 3
Europe, 106, 185
European Commission's Joint Research
    Center (JRC), 190
European Media Monitor (EMM), 191
Evacuation, 137, 141, 162,
    165–170, 175, 189, 194, 200,
    206–207
Evaluation, 12, 32, 44, 83, 85–87,
    89–90, 92, 95, 132, 200
Events data, 2–4, 7, 12–13, 15, 17–22,
    24–25, 34–36, 39–52, 78, 89, 91–92,
    94–99, 101–103, 105–110, 112–113,
    115–116, 118–120, 124–125,
    127–128, 131, 133, 135–143,
    145–148, 182, 184–185, 189,
    191–193, 195–197, 199, 201,
    203–204, 206
Exaggeration of threat, 18–19, 36,
    166, 204
Expert systems, 194–195
Extremists, 22, 24–25, 28, 30–32, 114,
    149, 151, 171, 186

Facebook, 1, 108–109
Farsi, 193
Felcan, David, 196
Fernando, Gayathri, 152
Festinger, Leon, 30
Finland, 186, 188
Fischer, Peter, 197
Fisher, Roger, 157
Flake, Gary, 194
Ford, David, 11

Forecasting of Crises and Instability
  using Text–based Events
  (FORECITE, 184
Forgiveness, 73, 158
Foundation for Co-Existence (FCE), 3,
  12–13, 34, 41, 46, 92–99, 101–103,
  106, 113, 116, 118, 127, 129,
  138–140, 142–143, 148–149,
  154–155, 182, 184–185, 191, 195,
  204
Free Burma Rangers (FBR), 170–171
Free-rider problem, 197
Freedom Fone, 191
Freire, Paulo, 31, 118
FrontlineSMS, 90
Fundamental indicators, 38–39, 52,
  131. *See also* Technical indicators
Fundamentalism, 75, 171
Funders, 14, 79, 93, 129, 140, 177, 189
Fuzzy Analysis of Statistical Evidence
  (FASE), 39
Fuzzy logic, 42, 195

Gaasbeek, Timmo, 46, 95–97, 142, 148,
  185
Gandhi, Mohandas (also known as
  Gandhiji), 8, 58, 69, 123, 151–152
García, Alejandro, 158–159
Garner, Karen, 5
Gatekeeping, 29
Gaza Strip, West Bank, 108, 167
Gender relations, 41, 62, 87, 162
Genocide, 5, 41, 174, 185, 188
Geocoding of events data, 106, 124,
  184
Geographic information system (GIS),
  18, 41, 105, 107, 120, 124, 136,
  193–194
George, A. L., 11
German Agency for Technical
  Cooperation (GTZ), 129
Germany, 60
Gerner, Deborah, 40, 43–44, 46, 131

Ghana, 108, 165
Gilligan, James, 81
Giridharadas, Anand, 105
Global Positioning System (GPS), 120,
  126, 194
Goa, India, 77
Gohar, Ali, 152–153
Goldfinch, Shaun, 11
Goldstone, Jack, 11
Gomtipur, India, 72
Goodwill, as in between identity
  groups, 58, 180, 199
Google, 114
GoogleEarth, 20, 106–107
Gopin, Marc, 27
Governance, 9, 40, 91, 93, 205
Government of Sri Lanka (GoSL),
  21, 154
Gowing, Nik, 105
Grameen Bank, 6
Graphic depiction of information, 10,
  45, 49, 91, 115, 137, 146
Great Britain, 142, 163
Greed-grievance nexus, 24–25, 35–36,
  51, 59, 65–66, 124, 204
Guarino, Mark, 82
Guerrillas, 115, 158, 168, 186
Guinea, 165
Gujarat, India, 57, 60, 64–65, 77
Gujer, Eric, 142
Gurr, Ted Robert, 11, 17, 25

Haiti, 11, 108, 110, 119–122, 125, 164,
  184, 205
Hansen, Silke, 157
Hardin, Russell, 197
Harff, Barbara, 25
Harry S Truman Institute, Hebrew
  University, 148
Hartel, 41
Hartnett, Susan M., 85
Harvard Humanitarian Initiative, 5, 21,
  119, 193

Harvard University, 5, 119, 157, 193
Hate speech, 141, 184, 191
Hauben, Michael, 117
Hauben, Ronda, 117
Hebrew University, 148
Hegre, Haavard, 8, 10
Heinzelman, Jessica, 108, 110, 113,
    122, 184
Hermeneutic variability, 27, 31–32, 36,
    151, 204
Herrmann, Roy, 165
Hersman, Erik, 6, 109, 113
Hess, Gregory D., 50
Hewlett Foundation, 91
Hibernate, as compared to evacuate,
    162, 165, 168, 175, 194, 200, 207
Hidden Markov models (HMM), 12,
    50–51, 53, 142
High-frequency radio (HF), 13, 127,
    132, 134
High gain antenna phones, 13
Himelfarb, Sheldon, 193
Hindus, 12, 30, 49, 58–61, 65, 67–70,
    72–75, 77, 92–94, 148, 151–152,
    155, 185
Hippocratic oath, 164
Histogram, 99
Hobbs, Jerry R., 42–43
Hoeffler, Anke, 24
Hoffman, Ben, 5
Hoffman, Evan, 5
Holl, J. E., 4, 11
Holland, R., 31
Holocaust, 5, 27
Horowitz, Donald, 18–22, 33–35, 46,
    75, 79, 99–101, 103, 143, 150
Howard, Philip, 1
Human coded events data, 44
Humanitarian Accountability
    Partnership (HAP), 200
Hunchlab (Crime Spike Detector), 142,
    189, 196
Hussein, Saddam, 166

Hussein, son of Ali, 67–68
Hutus, 41, 47, 110, 186
Hypergeometric approach, 196
Hypotheses, 17, 34, 110, 123, 205

Iceland, 186, 188
Imperfect information, 146. See also
    Uncertainty
Improvised explosive device (IED), 172
Inaccuracy, 31, 44, 52, 144
Incident reports, 128, 131, 144, 191
Incitement, 41, 51, 66, 108–109, 147,
    149, 151, 156, 158, 191
India, 6, 12, 22, 24, 49, 57–58, 60, 69,
    71–72, 75, 77, 108–110, 113,
    151–153, 157–158
Indonesia, 91
Inductive reasoning, 13, 18, 46, 48–49,
    72, 79, 96–98, 102, 135, 140–142,
    195–196, 205
Inflammatory rhetoric or propaganda,
    68, 71, 109, 156
Information and communication
    technologies (ICTs), 1–2, 115, 161
Information fire hose, 112
Iniquity theory, 25, 35
Inoculate, as against communalism,
    65, 79
Integrated Regional Information
    Networks (IRIN), 122
Inter-Governmental Authority on
    Development (IGAD), 3, 13, 127–
    129, 133
Interest-based negotiation, 157
Interfaith relations, 28–30, 71,
    73, 172
Internally displaced people (IDPs), 162,
    164, 168, 170
International Federation of Red Cross
    and Red Crescent Societies, 162
International non-governmental
    organizations (INGOs), 5, 57, 94,
    102, 165, 168, 199, 206–207

Internet, 13, 17, 47, 56, 78, 89, 97, 106, 108, 110, 115–116, 121, 125, 136–138, 141, 144, 149, 177, 184, 189–190, 192–195, 199
Internet Bar Organization (IBO), 193–194
Internews, 144
Interrogation of data, 189, 195–196, 205
Intervention, 7, 10–15, 17, 21–23, 33–34, 36, 46, 71–72, 78–79, 86, 89–90, 95, 98, 101, 103, 109, 117, 125, 128–129, 131–132, 140, 147–148, 153, 155, 159, 161, 163, 170, 174–175, 184–185, 188–189, 197, 199, 204–205
Intrafaith relations, 28–31, 148. *See also* Coreligionist communication in anti-incitement
Iran, 1, 114, 142, 190
Iraq, 85, 166–167, 172
Ishmael, 174
Ishwar, 64
Islam, 85, 124, 148–149, 151
Israel, 27, 166
Iyer, Pushpa, 6

Jains, 58, 60
Jamaica, 85
Janjaweed militia in Sudan, 5
Japan, 108
Jenkins, J. Craig, 41
Jerusalem, 148–149, 159, 166–167
Jews, 148–149
Johansen, Robert, 151
Jordan, 44

Kakar, Sudhir, 22
Kane, Candice, 6, 82
Kansas City, Missouri, 85
Kansas Events Data System (KEDS), 43
Kansas, University of, 43
Karachi, Pakistan, 110, 153

Karamoja Cluster, East Africa, 128–129, 131
Karen, ethnic group of Burma (Myanmar), 170–171, 174
Kashmir, disputed territory between India and Pakistan, 58
Kathmandu, Nepal, 91
Kenya, 6, 13, 20, 32, 48, 90, 105, 107–109, 111, 113, 115–117, 119, 121, 123, 125, 127, 129–132, 159, 181
Khalil, Abdelrahim Ahmad, 6
Khan, Akhter Hameed, 153
Khan, Ghaffar, 151–152
Khudai Khidmatgars (translated to mean "Servants of God"), 151–152
Kibaki, Mwai, 20
Kikuyu tribe in Kenya, 20
King, Martin Luther, Jr., 123
Kobia, David, 6
Korf, Benedikt, 24, 27
Kosovo, 28, 32
Kreps, Sarah, 5
Kuroda, Michiko, 11
Kurtz, C., 39
Kuse, Doreen, 107
Kuwait, 172
Kyrgyzstan, 85

Lacey, Marc, 181
Lancaster, Kevin, 146
Land, Molly Beutz, 117
Latané, Bibb, 197
Lawrence, Patricia, 46, 95–97, 142, 148, 185
Lederach, John Paul, 7, 21–22, 33, 155–159
Levels of leadership, 12–14, 22–23, 55–56, 123, 151, 177–178
Lexical processing, 42–43
Liberation Tigers of Tamil Eelam (LTTE), 8, 21, 43, 92–93, 154–155, 186, 188

Liberia, 116, 165
Libya, 108, 119, 123
Lieberman, Marya, 193
Lipsey, R. G., 146
Local area networks (LANs), 47
Locally led advance mobile aid, 162
Logan, B. I., 105
Louisiana, 85
Lowe, Will, 42–45
Lulls, as occurring between
  precipitating events and violence,
  18–24, 34, 36, 98–103, 138, 140, 143,
  203–204
Lund, Michael, 4, 9
Luxembourg, 60
Lyke, Susan, 154

Maasen, Kristel, 161
Mabey, Nick, 39–40, 183
Magdalena Medio, Colombia, 157, 160
Mahajan-no-Vando slum of
  Ahmedabad, India, 58, 61–62, 72, 74
Maharashtra, India, 77
Malawi, 32, 108
Malone, D., 24
Malone, Thomas, 55
Manos Unidas, 60
Mapping, 5, 13, 48, 89, 107–108, 112,
  116, 119–121, 123, 136, 184, 193,
  195
Maryland, 85
Mashada, 107
Mashup, 107
Massachusetts, 119
Massachusetts Institute of Technology
  (MIT), 55, 111
Mathematics, 15, 39, 46, 51–52, 106,
  112, 125, 127, 136, 140–141, 143,
  191
Matveeva, Anna, 2
Maysan Governorate, Iraq, 172
McCauley, Clark, 17, 19, 25, 31, 147
McClelland, Charles, 44

McCombs, M.E., 73
Meadows, Donella, 55
Mecca, Saudi Arabia, 148
Media Focus on Africa Foundation
  (MFOA), 115
Mediation, 17, 60, 63, 70, 77, 83, 94,
  102, 140, 149, 155, 157, 165, 186,
  207
Meertens, R. M., 31
Meier, Patrick, 4, 40, 45, 81, 98, 106,
  108, 111–112, 114–116, 127, 133,
  143–146, 182, 192
Mercy Corps, 162
Merton, R. K., 17
Messiah, 149
Methodologies, 2, 6–8, 13–15, 18, 44,
  47–49, 53, 101, 106, 113, 124, 133,
  135–136, 140, 153, 174, 190, 204,
  206
Mexico, 57, 85
Micro Level Analysis of Violent
  Conflict (MICROCON), 11
Microchips, 132–133
Microfinance, 6
Misereor, 60
Missouri, 85
Mobile data computers (MDCs), 142
Mohajirs, 153
Mohammad, the prophet, 6
Molotov cocktail, 110
Momentum, 25, 45, 84
Moral dissonance, 31. See also
  Cognitive dissonance theory
Morozov, Evgeny V., 190
Moscow, Russia, 3
Moseley, W. G., 105
Mount Moriah (Temple Mount,
  Jerusalem), 148
Moving average convergence
  divergence (MACD), 50, 131
Moynihan, Colleen, 89
Mubarak, Hosni, 123
Muftis, 26–27, 76

Muhumuza, Joseph, 129
Mujahedeen, 150
Mullah, 150
Multan, Pakistan, 163
Mumbai, India, 57, 113–114
Munro, Robert, 121
Murshed, Syed Mansoob, 39
Music, 30, 65
Muttur, Sri Lanka, 149, 154
Mwaûra, Cirû, 128
Myanmar, 170

Nairobi, Kenya, 13, 48, 89, 181
Nanotechnology, 5
National Gang Intelligence Center, 82
National Institute of Urban Affairs, 58
National research institutes (NRIs), 129
Nefarious actors, 8, 113–114
Negotiation, 8, 149, 155–157, 164, 172,
    207. *See also* Mediation
Nepal, 91
Netherlands, The, 60, 95, 108
Netizen democracy, 117. *See also*
    Democratization
Network for Ethnic Monitoring and
    Early Warning (EAWARN), 3
Niger Delta, Nigeria, 91
Nominal data, as compared to interval
    data, 99
Non-governmental organizations
    (NGOs), 6, 23, 40, 48–49, 52,
    57, 59–61, 76–77, 79, 94, 96–97,
    102, 107, 129, 134, 137, 144,
    153, 162–163, 165, 169–170,
    172–173, 181, 184, 189–190,
    199, 206–207
Nonfood items, 121
Norms, 33, 35, 84, 86, 90, 132, 147,
    151–152, 159, 203, 206
Northwestern University, 32, 85, 193
Norway, 93, 186, 188
Notebook computers, 2, 189
Notre Dame, University of, 144, 193

Nyheim, David, 3–4, 42, 55, 91, 179,
    182–183, 192

O'Brien, Sean, 39–40, 51
O'Callaghan, Sorcha, 165
Odinga, Raila, 20
Office for the Coordination of
    Humanitarian Affairs (OCHA),
    United Nations, 119
Oh, Churl, 41
Okolloh, Ory, 6, 107
Olson, Mancur, 197
Operational code, 51, 53
Operational violence prevention, 180,
    187, 204
Options, 18, 21–22, 33, 36, 147, 152,
    155, 157–160, 165, 204, 206
Orangi Pilot Project (OPP), 153–154
Organisation for Economic
    Co-operation and Development
    (OECD), 3
Organization of Security and
    Co-operation in Europe (OSCE), 185
Orthodox religion, 28, 30, 32, 149
Oscillations of tensions, 10–11, 203
Oslo, Norway, 106
Outsiders, 3, 5–6, 9, 79, 97, 121, 162,
    185, 188, 197, 204, 207
Oxfam/Great Britain, 163

Pacifists, 156
Pakistan, 26, 37, 58, 71, 75, 91, 110,
    151, 153, 159, 162–163, 167
Palestine, 166
Palestinians, 43, 148–149, 167
Palihapitiya, Madhawa, 6, 154
Pantuliano, Sara, 165
Paradi-Guilford, Cecilia, 193
Paradox of early warning, 166
Parsing software, 43, 192–193
Pastoralists, 3, 13, 128, 205
Patel, Pravin, 59
Patel, Ricken, 117

Pattan, 163
Pattern recognition, 12–13, 37–39, 46, 49, 51, 53, 96–98, 102, 106, 127, 136, 138–143, 182, 190, 194, 205
Payson Conflict Study Group, 25
Peace agreements, 8, 92, 94–95, 154, 164, 186, 188
Peace Research Institute of Oslo (PRIO), 106
Peacebuilding, 7–8, 115, 133, 182–183, 203
Peacekeepers, 22–23, 36, 70, 164, 72, 181, 185–186, 188, 204
PeaceTXT, 90
Pennsylvania, 85, 196
Pennsylvania, University of, 196
Pentagon, 47
Perera, Robert, 98
Persuasion, 156, 207
Peshawar, Pakistan, 152
Philadelphia Police Department, 20, 142, 189, 196
Philippines, The, 162
Photojournalism, 92
Piza-Lopez, Eugenia, 42
Pizza delivery, example of early warning data, 47
Platforms, 48, 90, 107–108, 113–114, 117, 119–120, 122–123, 126, 190
Playwrights, 69
Podcasts, 1
Poetry, 64
PopTech, 90
Posters, 64, 71, 84
Prabhakaran, Velupillai, 8
Prakash, Cedric, 6, 31, 57–58, 63–64, 70–72, 74–77, 95
Preaching, 30, 81
Precipitating events, 19–24, 34–36, 39, 99, 101, 148, 197, 203–204. See also Provocative events
Predictive tagging, 112

Preempting violence, 22, 63, 70–71, 180, 187
Prendergast, John, 5
Preparedness, 2, 81, 156, 162–163, 165–166, 170, 173–174, 200–201, 206–207
Prepositioned supplies, 163–164
Prestige, as an incentive for violence, 8, 132
Prevention of violence, 2, 5–14, 17, 19, 21–23, 25, 27–29, 31, 33–35, 37, 41–42, 52, 55, 57, 63, 70, 72, 76–79, 81–84, 89–91, 95, 98, 105, 116–117, 119–120, 122–123, 125, 129, 131–135, 137, 139, 143, 147, 159, 161, 163, 165, 167, 169, 171, 173, 175, 177, 179–180, 182–184, 187, 189, 191, 193–194, 197–201, 203–207
Priest, 32, 150, 185
Probes, 32, 39–40
Promotive approaches to fostering intergroup goodwill, 63, 70, 79
Propaganda, 59, 71, 75, 115, 157
Protection, 63, 73, 77, 161–165, 170–171, 173, 179
Protesters, 2, 116, 123, 190
Provocative events, 7, 98–99, 101, 103, 141, 156, 190, 201. See also Precipitating events
Pukhtoons, 152–153
Punjab, Pakistan, 26

Quakers, 156
Qur'an, 26

Rabbi, 149, 159
Radio frequency identification (RFID), 127, 132–134
Raids, 13, 128, 133, 205. See also Cattle rustling
Rainfall, 45, 105, 163
Rallies, 51, 84, 117–118

Ramadan, 49, 149
Randomization, 101, 123
Rapid Response Fund (RRF), 129
Rapoport, David C., 27, 29
Reader, of Virtual Research Associates,
    44–45, 191–192
Reality score, 112
Really Simple Syndication (RSS), 108
Rebellion, 34, 43, 92–93, 108, 117, 174
Reconciliation, 29, 154, 206
Reconstruction, 10, 153–154, 183
Referendum, 124
Refugees, 152, 162, 164–165, 168–169,
    172, 199
Relative deprivation, 17, 25
Religious illiteracy, 30
Renegade militant groups, 111, 114,
    185, 188
Response reports, 131
Restricted feeding, 14, 118, 122,
    124–126, 198. See also Bounded
    crowd feeding
Reuters, 43–44
Rigged elections, 19–20
Riots, 18–20, 22, 26, 28, 34, 39, 58–59,
    61–62, 69, 73, 75, 77, 107, 110, 122,
    142, 150–153, 192
Risk analysis, as compared to early
    warning, 12, 20
Ritter, Nancy, 83
Robert Woods Johnson Foundation, 85
Rotich, Juliana, 6
Rudge, David, 149
Rudolph, Lloyd I., 18
Rudolph, Susan H., 18
Rumors, 19, 28, 71, 115, 159
Rupesinghe, Kumar, 6, 11, 94–95
Rwanda, 11, 41, 185, 188

Sahajeevana Centre for Coexistence, Sri
    Lanka, 92
Salazar, Oscar, 109
Satellite Sentinel Project, 5

Satellites, 5, 172, 190
Saudi Arabia, 148
Scandinavian countries, 94, 154
Scenarios, 20, 89, 151, 156
Schmeidl, Susanne, 42, 128
Schön, Donald, 35
SchoolTipline, 85
Schoonman, Marten, 115
Schrodt, Philip, 40, 43–44, 46, 50, 131
SCUD missile, 166–167
Second best, theory of, 146, 206
Senge, Peter, 169
Senn, Dominic, 41
Serbian, 28, 32
Servants of God, 151–152, 159
Settlers, 92, 130
Sextant, 194
Shahin, Emad El-Din, 124
Shareware, 13
Sharp, Gene, 122
Shaw, D. L., 73
Shaykhutdinov, Renat, 24
Shelat, K.N., 59, 71
Shiite, 67
Shootings, 82, 86–87
Short Message Service (SMS), 82, 89,
    107, 114, 131, 138, 194
Short-circuiting proposition, 21
Shrine, 75, 149
Sierra Leone, 165
Sikhs, 58, 152
Silk Road Initiative, 193–194
Simon, Herbert, 130
Simulation, 151, 180, 189, 194
Sinhalese, 8, 92, 94–96, 149, 154–155,
    159
Siror, Joseph K., 132
Situation reports, 96–97, 122, 128, 131
Situational awareness, 122, 125, 156
Skogan, Wesley G., 32–33, 85–87
Skype, 107
Slutkin, Gary, 6, 32, 81
Smith, Tony, 196

Snowden, D., 39
Social media, 1–2, 12–13, 40, 81, 108, 114, 122, 124, 136, 138, 141, 144, 177, 195, 199, 205
Solomon, King, 148
Somalia, 11, 13, 32, 127, 129, 162, 181
Songs, 64–66, 68–69, 73, 79
Sons of the soil conflicts, 27, 35
Soto, Jorge, 109
South Kivu, Democratic Republic of Congo, 164
Spain, 60
Spatial dimension, 107, 196
Spoilers, 8, 10, 23, 31, 36, 110, 124–125, 149, 180, 187, 204. *See also* Troublemakers
Spoilers to stakeholders strategy, 10
Spousal abuse, 60, 63
Spyglass, 194
Sri Lanka, 3, 6, 8–9, 12, 21, 27, 34, 41, 43, 51, 91–95, 97, 99–103, 139–140, 154–155, 159, 180, 182, 186, 188
Sri Lanka Army (SLA), 154
Sri Lanka Monitoring Mission (SLMM), 94, 98, 102, 154–155, 186, 188
St. Xavier's Social Services Society, 12, 31, 59–65, 69–79, 84, 90, 96, 116, 118, 129, 138–139, 143, 150, 153, 155, 204
Stakeholders, 10, 130, 161
Stanford University, 184
Statistics, 3, 39–40, 47, 49, 86, 92, 98–99, 103, 131, 133, 144, 196
Steele, David, 31–32
Stephan, Maria, 123
Stewart, Frances, 25
Stickers, use of in raising awareness, 84
Stockholm International Peace Research Institute (SIPRI), 40
Strategic nonviolence, 7–8, 122, 199, 205
Strategic peacebuilding, 7–8, 203

Strategies, 7, 10, 13, 63, 71, 74–78, 86, 113, 115, 122–123, 131–133, 147, 159, 161, 170, 183, 199, 205
Students, 44, 64, 85, 111, 116, 149, 151–152, 184, 193, 196
Sudan, 5, 13, 85, 127, 129, 164
Sunstein, Cass, 106, 141
Surowiecki, James, 141
Surveillance, 5, 191
Swahili, 107
Sweden, 60, 186, 188
swisspeace, 3, 40–41, 47, 94, 102, 140, 183
Switzerland, 106
Symbols, 10, 19, 27, 69, 72, 74–76, 99, 193, 204
Syntactic processing, 42–43

Tablets, 163
Tactics, 7–10, 30, 70, 86, 133, 137, 163–165, 170–171, 175, 180, 199, 205
Taliban, 152
Tamils, 8–9, 12, 21, 43, 92–96, 148–149, 154–155, 159, 184–186, 188
Tanzania, 32, 165
Taylor, Charles Lewis, 41
Technical indicators, 38, 52, 131, 191. *See also* Fundamental indicators
Tehran, Iran, 142, 192
Telecommunications, 131, 165
Television, 2, 70, 95, 114
Templeton College, Oxford University, 47
Temporal dimension, 19, 107, 196
Tennekoon, Brigadier General Sunil, 21–22
Terrorism, 113, 117, 182
Thailand, 32, 91
Theatrical performance, 19
Theodore, Jesse, 196
Theological dueling, 31

Theology, 27, 30–31, 151
Theory, 5, 12–13, 15, 17–19, 21, 23, 25, 27, 29–33, 35, 55, 85–86, 90, 146–147, 151, 160, 186, 197, 203–204, 206
Threats, 4, 18–19, 21, 36–37, 41, 46, 53, 66, 72–73, 75, 77, 120, 137, 147, 152–156, 159, 161, 165–167, 170–171, 174, 191, 200–201, 204, 206–207
Tiba Colony, Pakistan, 26–27
Timeliness, 15, 75, 79, 89–90, 118, 130–132, 135, 143–144, 146, 155, 175, 178, 195
Tito, Josip Broz, 32
Tobago, 85
Toft, Monica Duffy, 27
Tokenization, 42–43
Tolerance, 26, 84, 174. See also Diversity, celebration of
Torture, 9, 93, 173
Training, 70, 82, 89, 129. See also Workshops
Transistor radio, 163
Translation, 135, 190, 192
Transnational organizations, 4, 6–7, 162–163, 185, 192
Trellon, LLC, 5
Tribal conflict, 2, 13, 20, 127–129, 131, 133, 152–153
Trincomalee, Sri Lanka, 21, 93, 98, 149, 154–155
Trinidad, 85
Troublemakers, 2, 8, 23, 31, 36, 110, 124–125, 148, 180, 187, 204. See also Spoilers
Trust network, 113, 119, 122–123, 126, 198–199, 205–206
Tsunami, 41, 93, 154, 182
Tufts University, 184
Tully, Melissa, 115
Tunisia, 1, 122, 124, 198, 205
Tutsi, 41, 47, 110, 186
Twitter, 1, 93, 108–109, 190

Uganda, 13, 32, 127, 129, 132
Ullberg, Lars, 89
Uncertainty, 35, 168, 178, 207. See also Imperfect information
United Kingdom, 11, 184
United Nations, 5, 119, 122, 164–165, 168–170, 172, 181, 185–186, 199, 207
United Nations High Commissioner for Refugees (UNHCR), 165
United Nations Institute for Training and Research (UNITAR), 5
United Nations Mission to the Democratic Republic of the Congo (UN MONOC), 164
United States Agency for International Development (USAID), 5, 129, 166–167
Unmanned aerial vehicles, 5
Ury, William, 157
Ushahidi, 20, 48, 89–90, 105, 107–116, 118–125, 136–139, 143–144, 184, 193, 195, 205

Validation of data, 38, 46, 48, 95, 98–99, 102, 106, 110–113, 116, 118, 125, 136, 138, 143–146, 156, 166, 190–191, 194, 199, 206
van Vugt, M., 31
Varshney, Ashutosh, 19
Venice, Italy, 38
Victims, 9, 15, 24, 30, 41, 61, 70, 73, 83, 85, 90, 93, 141, 143, 154, 161, 198
Videosharing, 1
Videotaping, 93
Virtual Research Associates (VRA), 42, 44, 96, 131, 191–192
Visualization, 139, 195–196, 203, 205
Voix des Kivus, 5
Volunteer and technical communities (V&TCs), 107
Voters, 109–110

W. K. Kellogg Foundation, 57
Wageningen University, 95
Warning-response problem, 4–6
Warnings, 4, 6, 11–12, 14–15, 21, 37,
    45, 48, 51–52, 79, 96, 98, 103, 108,
    113, 116, 118, 122, 125, 128, 131,
    135–140, 142–143, 163, 185, 188,
    191, 194–195, 206–207
WarViews, 106–107
Waters, Carol, 108, 110, 113, 122, 184
Watkins, Michael, 37
Weidmann, Nils B., 107
Weinberg, Tamar, 1
Wigger, Andreas, 141, 161
Wikipedia, 1, 107–108, 111, 118, 125
Wohlgelernter, Elli, 149
Women's Commission for Refugee
    Women and Children, 165
Workshops, 41–42, 70, 77, 94, 96.
    See also Training
World Bank, 8
World Events Interaction Survey
    (WEIS), 43–44
Wulf, Herbert, 130, 142

Xenophobic religious nationalism, 31,
    76

Yahoo, 191
Yiu, Chris, 39–40, 183
Youth, 49, 74, 83–84, 87, 89, 93, 111,
    148–149, 154, 159
YouTube, 111
Yugoslavia, 32
Yunus, Mohammad, 6

Zadeh, L. A., 43
Zaire, 32, 162
Zarrella, Dan, 111
Zealotry, 149, 184, 187
Zimbabwe, 191
Zoroastrians, 58